Fumigation for Insect Control

Fumigation for Insect Control

Editor

Dr. K.R. Nair

2014

Gene-Tech Books

New Delhi - 110 002

ISBN 81-89729-78-0

Published by : **GENE-TECH BOOKS**
4762-63/23, Ansari Road, Darya Ganj,
NEW DELHI - 110 002
Phone: 41562849
e-mail: genetechbooks@yahoo.co.in

Printed at : **Tarun Offset Printers**
Delhi

PRINTED IN INDIA

Preface

Fumigants are widely used for the control of insects and other pest organisms. Because of their unique characteristics and the great adaptability of the fumigation technique, fumigants can often provide effective, economical control where other forms of pest control are not feasible. In many cases treatments can be carried out on infested material without disturbing it in any way. Several modern technological developments, including instrumentation for gas detection and analysis, improved formulations as well as increased demand for effective and economical pest control measures, have done much to improve fumigation procedures.

Modern technology and research have also brought to light certain problems with fumigants that were previously unknown. Numerous investigations made on both the acute and chronic effects of fumigants have shown that some of these materials are capable of producing serious effects on human health. In some cases fumigants with excessive hazard potential have been restricted or prohibited so that they are no longer widely used for pest control in some countries. For the effective use of fumigants in pest control, personnel assigned to fumigation work should receive thorough instructions on the properties of fumigants and training in safe methods of handling.

This book deals with fumigation for the control of insects above the ground. It is written for practical fumigators officials who is required to conduct or supervise fumigation treatments. The book may also be of interest to planners and consultants in crop protection who need information on the scope and limitations of fumigation as an instrument for insect control.

Editor

Contents

1

Introduction

Fumigation is a method of pest control that completely fills an area with gaseous pesticides to suffocate or poison the pests within. It is utilised for control of pests in buildings (structural fumigation), soil, grain, and produce, and is also used during processing of goods to be imported or exported to prevent transfer of exotic organisms.

Structural fumigating techniques differ from building to building, but in houses a rubber tent is often placed over the entire house while the pesticides are being released into the residence. This concentrates the gases as well as keeps them from escaping and doing harm to people in the neighbourhood. During this time the residents of the house must find an alternate residence for up to a week depending on the fumigmant used, which in turn depend on the severity of infestation and size of the house.

Widely used fumigants include:

- phosphine
- 1,3-dichloropropene
- chloropicrin
- methyl isothiocyanate

— hydrogen cyanide

— sulfuryl fluoride

Fumigation is a hazardous operation, and generally it is a legal requirement that the operator carrying out the fumigation operation holds official certifcation to perform the fumigation. Fumigation usually involves the follwing phases. First the area to be fumigated is usually covered to create a sealed environment; next the fumigant is released into the "space" to be fumigated; then, the "space" is held for a "set" period while the fumigant gas percolates through the space and acts on and kills any infestation in the product, next the "space" is ventilated so that the poison gasses are allowed to escape from the space, and render the "space" safe for humans to enter.

The correct ventilation of the area is a critical safey aspect of ventilation. It is important to distinguish between the pack or source of the fumigant gas and the environment which has been fumigated. While the fumigant pack may be safe and spent, the "space" will still hold the fumigant gas until it has been ventilated.

There are many chemical compounds which are volatile at ordinary temperatures and sufficiently toxic to fall within the definition of fumigants. Highly flammable compounds are not necessarily excluded if dangers of fire and explosion can be controlled by the addition of other suitable compounds, or if fumigation procedures are carefully designed to eliminate these hazards. Toxicity to human beings is not necessarily a cause for exclusion. All known fumigants are toxic to humans to a greater or lesser degree and ways can be devised for their safe handling under the required conditions of application.

However, some commonly used compounds have been shown to be capable of producing long-term effects that were previously unknown. The use of such fumigants is becoming more restricted and some materials have already been eliminated from the list of fumigants approved for use in certain countries. Table 1 lists a number of common insecticidal fumigants which have been used for many years.

The final selection of a fumigant for any particular pest control problem will be influenced by the various properties of the compound along with the type of pest organism and the nature of the commodity. As the number of chemicals approved for use as fumigants is small, and is declining, this selection is being narrowed to a very few.

The boiling point of different chemical compounds generally rises with the increase of molecular weights. This generalisation is borne out by the data for the fumigants shown in Figure 1, where molecular weights are plotted against boiling points. The relationship stated above holds very well, except for methyl bromide, and it demonstrates that important compounds, such as carbon tetrachloride or ethylene dibromide, will evaporate very slowly under practical fumigation conditions.

If the highest possible concentrations are required at the beginning of the fumigation with such compounds, more rapid volatilsation will have to be effected in some way. Figure 1 shows that, from the physical standpoint, fumigants may be divided into two main groups according to whether they boil above or below room or moderate outdoor temperatures (20°C to 25°C). The low boiling point fumigants, such as methyl bromide, may be referred to as gaseous -type fumigants. These are kept in cylinders or cans designed to withstand the pressure exerted by the gas at the highest indoor or outdoor temperatures likely to be encountered.

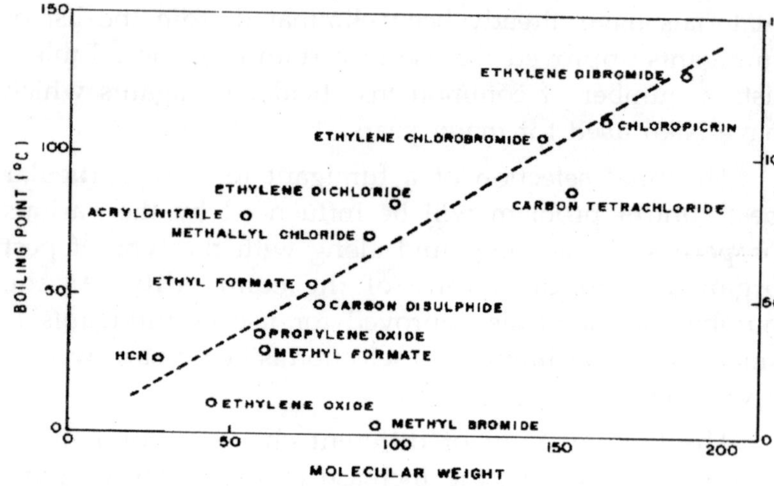

Figure 1. Relationship between molecular weight and boiling point of some fumigants.

The second main group of fumigants contains those with high boiling points; these are usually described as liquid-type or solid-type according to the form in which they are shipped and handled. In some kinds of work, such as grain and soil fumigation, the slow evaporation of certain liquids is an advantage because the initial flow leads to a better distribution of the gas subsequently volatilised. In other applications, where personnel have to distribute the fumigants, slow evaporation of the liquids or solids makes them safer to handle.

Included in the general term solid-type fumigants are certain materials which are not fumigants themselves, but which react to form fumigants after application. Examples are calcium cyanide powder, which reacts with atmospheric moisture to yield hydrogen cyanide (HCN), and formulations of aluminium and magnesium phosphides which also react with moisture to produce

Table 1. Essential Properties of Fumigants in Common use for Insect Control

Name and formula	*Molecular weight*	*Boiling point °C*	*Solubility in water g/100 ml*	*Flammability Percent*
Acrylonitrile $CH_2 : CH.CN$	53.06	77.0	7.5 at 25°C	3-17
Carbon disulphide CS_2	76.13	46.3	0.22 at 22°C	1.25-44
Carbon tetrachloride CCl_4	153.84	77.0	0.08 at 20°C	Nonflammable
Chloropicrin $CCl_3.NO_2$	164.39	112.0	Insoluble at 20°C	Nonflammable
Dichlorvos (DDVP) $CCl_2{=}CHO.Po.(OCH_3)_2$	221	120°C/14mm	Slight	Nonflammable
Ethylene dibromide $CH_2Br.CH_2Br$	187.88	131.0	0.43 at 30°C	Nonflammable
Ethylene dichloride $CH_2Cl.CH_2Cl$	98.97	83.0	0.87 at 20°C	6-16
Ethylene oxide $CH_2.O.CH_2$	44.05	10.7	Very soluble at 20°C	3-80
Ethyl formate $H.COO\ C_2H_5$	74.05	54.0	11.8 at 25°C	2.7 - 13.5
Hydrogen cyanide HCN	27.03	26.0	Very soluble at 20°C	6-41
Methyl bromide CH_3Br	94.95	3.6	1.3 at 25°C	Nonflammable
Methyl formate $H.COO\ CH_3$	60.03	31.0	30.4	5.9 - 20
Paradichlorobenzene $C_2H_4Cl_2$	147.01	173.0	0.008 at 25°C	(Flash point***
Phosphine PH_3	34.04	-87.4	Very slightly soluble	1.79
Sulphuryl fluoride SO_2F_2	102.6	-55.2	Slight	Nonflammable
Trichloroethylene $CHCl : CCl_2$	131.4	86.7	Insoluble	Nonflammable

phosphine (hydrogen phosphide). There are also some fumigants in the form of crystals and flakes that sublime to give off fumigant vapours. Examples are paradichlorobenzene and naphthalene.

The maximum weight of a chemical that can exist as a gas in a given space is dependent on the molecular weight of that chemical. This fact, implicit in the well-known hypothesis of Avogadro, has an important practical application. It is useless attempting to volatilise in an empty chamber more fumigant than can exist in the vapour form. Table 2 shows the maximum amounts of a number of fumigants that can be vaporised in a given space. It will be noted that the fumigants with low boiling points, such as methyl bromide or ethylene oxide, may be released in large amounts compared with high boiling point compounds, such as naphthalene and paradichlorobenzene.

Sorption of the fumigant by the material treated in a given space will permit greater amounts to be volatilised. Nevertheless, the figures given will still apply to the amount which can exist as vapour in the free air space surrounding the fumigated material. Unless it is sustained by warming from an outside source, the temperature of an evaporating liquid constantly drops owing to the fall in energy caused by the escape of molecules with greater than average energy. Thus, evaporation takes place at the expense of the total heat energy of the liquid.

The number of calories lost in the formation of one gramme of vapour is called the latent heat of vaporisation of the liquid. Some fumigants have higher latent heats than others. Both HEN and ethylene oxide, with latent heats of 210 and 139 respectively, absorb considerably more heat in passing from liquid to than do methyl

Table 2. Maximum weights of various fumigants which can exist in vapour from in an empty fumigation space at different temperatures

Fumigant	*Maximum weight, in grammes/cubic metre at indicated temperatures*							
	0°C	*5°C*	*10°C*	*15°C*	*20°C*	*25°C*	*30°C*	*35°C*
	(32°F)	(41°F)	(50°F)	(59°F)	(68°F)	(77°F)	(86°F)	(95°F)
Acrylonitrile	102.6	129.8	164.4	206.3	252.9	319.1	397.8	482.4
Carbon disulphide	568.1	701.1	843.7	1010.9	1297.2	1430.8	1740.9	2 096.3
Carbon tetrachloride	288.5	363.0	460.9	572.6	730.9	916.8	1145.4	1 398.5
Chloropicrin	57.8	79.5	108.7	139.5	179.5	220.6	277.8	358.7
Dichlorvos (DDVP)	0.02	0.03	0.05	0.08	0.13	0.21	0.32	0.48
Ethylene dibromide	38.5	54.1	63.7	83.5	112.8	141.2	173.6	214.7
Ethylene dichloride	133.4	173.7	223.7	282.0	350.1	430.3	537.1	668.2
Ethylene oxide	1331.5	1606.6	1854.5	1862.4	1830.4	1800.0	1771.2	1 740.8
Hydrogen cyanide	418.7	532.0	643.4	751.3	900.4	1072.2	1084.7	1 067.7
Methyl bromide	3839.3	4152.8	4079.4	4008.6	3940.2	3874.1	3810.1	3 748.3
Naphthalene	0.15	0.22	0.33	0.43	0.56	0.69	0.95	1.40
Paradichlorobenzene	0.69	1.61	2.49	3.18	5.14	7.89	11.64	17.56
Phosphine	1514.4	1487.2	1460.9	1435.5	1411.0	1387.4	1364.5	1 342.3
Sulphuryl fluoride	4546.0	4464.2	4385.3	4309.2	4235.7	4164.6	4095.9	4 029.4
Carbon dioxide	1959.8	1924.6	1890.6	1857.8	1826.1	1795.4	1765.8	1737.1

bromide and ethylene dibromide, with latent heats of 61 and 46 respectively.

The factor of latent heat is of important practical significance. The high pressure fumigants, such as HCN, ethylene oxide and methyl bromide, are usually kept under pressure in suitable cylinders or cans. On release into the atmosphere, volatilisation takes place rapidly and, unless the lost heat is restored, the temperature of the fumigant may fall below the boiling point and gas may cease to be evolved. Also, as the liquid changing to gas is led through metal pipes and tubes, or rubber tubing, the fall in temperature may freeze the fumigant in the lines and prevent its further passage. Fumigants that are liquids at normal temperatures and are volatilised from evaporating pans or vaporising nozzles may require a source of heat, such as a hot plate, in order that full concentrations may be achieved rapidly.

Diffusion and Penetration

Fumigants are used because they can form insecticidal concentrations: (a) within open structures or (b) inside commodities and in cracks and crevices into which other insecticides penetrate with difficulty or not at all. Hence, it is necessary to study the factors that influence the diffusion of gases in every part of a fumigation system. This study includes the behaviour of fumigants both in empty spaces and also in structures loaded with materials into which the gas is required to penetrate.

Graham's law of diffusion of gases states that the velocity of diffusion of a gas is inversely proportional to the square root of its density. Also, the densities of gases are proportional to their molecular weights. Therefore, a heavier gas, such as ethylene dibromide, will diffuse more slowly throughout an open space than a lighter one such as ethylene oxide. The rate of diffusion is also directly related to temperature, so that a given gas will diffuse more quickly in hot air than in cold air.

Many of the commonly used fumigants are heavier than air. A notable exception is hydrogen cyanide. If a gas heavier than air is introduced into a chamber filled with air and it is not agitated by fans or other means, it will sink to the bottom and form a layer below the air. The rate of mixing between the two layers may be very slow. For example, in a fumigation of the empty hold of a ship with the heavy gas methyl bromide where the fumigator had neglected to place a circulating fan, a sharp demarcation was observed between the lower half with the gas, where all of the insects were killed, and the upper part, where complete survival occurred. In good fumigation practice, settling or stratification will not be encountered if adequate provision is made to disperse the gas properly from the very beginning of the treatment.

Even distribution can be ensured by employing singly or in suitable combination: multiple gas inlets, fans or blowers and/or circulation by means of ducts and pipes. Contrary to popular belief, once a gas or number of gases heavier than air have been thoroughly mixed with the air in a space, settling out or stratification of the heavier components takes place very slowly; so slowly, in fact, that once a proper mixture with air has been secured, the problem of stratification of a heavier than-air fumigant is of no practical importance for the exposure periods commonly used in fumigation work.

It has already been suggested that distribution and penetration can be aided and hastened by the use of blowers and fans. Such propellers may work free in the structure or through a system of circulating ducts. These devices may also add greatly to the efficiency of the fumigation process by hastening the volatilisation of high boiling point liquids from evaporating pans and by preventing stratification of heavy gases. Also, a factor known as the Turtle effect has proved useful in the fumigation of certain materials susceptible to injury.

It was shown that rapid stirring by a centrifugal fan in a fumigation chamber at atmospheric pressure greatly hastened the attainment of uniform concentrations of methyl bromide in all parts of a load of early potatoes, so that the consignment was not overdosed at the outside of the packages or under dosed at the centre. In a four-hour exposure period, rapid stirring for one hour at the beginning of the treatment was, to all intents and purposes, as effective as continuous stirring for the whole time.

Sorption

A very important factor affecting the action of fumigants is the phenomenon known as sorption. Fortunately, for the purpose of understanding fumigation practice, it is possible to give a general account of the important factors concerned. In the relationship of gases to solids, sorption is the term used to describe the total uptake of gas resulting from the attraction and retention of the molecules by any solid material present in the system. Such action removes some of the molecules of the gas from the free space so that they are no longer able to diffuse freely throughout the system or to penetrate further into the interstices of the material.

In fumigation practices, collision with air molecules tends to slow down gaseous diffusion through the material and sorption takes place gradually. Thus, there is a progressive rather than immediate lowering of the concentrations of the gas in the free space. Throughout the exposure period of six hours, the fall in concentration of methyl bromide was proportionately less compared with that for the three other fumigants, both in the empty chamber and with the two loads of oranges. This was due to the fact that the internal surface of the chambers and the boxes of oranges both sorbed less of the methyl bromide than of the other gases in proportion to the applied dosage. Sorption under a given set of conditions determines the dosage to be applied, because the amount of fumigant used must be sufficient both to satisfy the total sorption during treatment and also to leave enough free gas to kill the pest organisms.

The general term sorption covers the phenomena of adsorption and absorption. These two are reversible because the forces involved, often referred to as van der Waal's forces, are weak. On the other hand, a stronger bonding called chemisorption usually results in chemical reaction between the gas and the material and is irreversible under ordinary circumstances.

Physical sorption

From the point of view of practical fumigation, adsorption and absorption, being both physical in nature and reversible. However, it is necessary to make some distinction between them at the outset because the forces involved may be less with adsorption than with absorption. Stated briefly, adsorption is said to occur when molecules of a gas remain attached to the surface of a material. Because some absorbents, such as charcoal or bone meal, are highly porous bodies with large internal surfaces, adsorption may also occur inside a given body. Absorption occurs when the gas enters the solid or liquid phase and is held by capillary forces that govern the properties of solutions.

For instance, a gas may be absorbed in the aqueous phase of grain or in the lipid phase of nuts, cheese or other fatty foods. Physical sorption, considered generally, is an extremely important Factor affecting the successful outcome of fumigations. Apart from specific reactions between certain gases and commodities, it may be stated as a general rule that those fumigants with higher boiling points tend to be more highly sorbed than the more volatile compounds.

Physical sorption varies inversely as the temperature, and is thus greater at lower temperatures. This fact has important practical applications. It is one of the reasons why dosages have to be progressively increased as the temperature of fumigation is lowered. Sorption may also be influenced by the moisture content of the commodity being fumigated. This was demonstrated by Lindgren and Vincent in the fumigation of a number of foodstuffs with methyl bromide; at higher moisture contents more fumigant was sorbed. This effect may be important with fumigants which are soluble in water to any significant degree.

The specific physical reaction between a given gas and a given commodity cannot be accurately predicted from known laws and generalisations. Usually, a certain fumigant must be tested with each material concerned before a recommendation for treatment can be drawn up.

Desorption

When a treatment is completed and the system is ventilated to remove the fumigant from the space and the material, the fumigant slowly diffuses from the material. This process is called Resorption and is the reverse of physical sorption. With the common fumigants and the commodities usually treated, residual vapours are completely dissipated within reasonable periods, although the length of time varies considerably according to the gas used and the material treated. Because of the inverse effect of temperature, dissipation of the fumigant usually takes place more slowly when the material is cold and may be hastened by warming the space and its contents.

Humidity also facilitates desorption of fumigants; at high humidity, wheat fumigated with ethylene dibromide was found to desorb 80 percent more of the fumigant than at very low humidity. As humidities can change appreciably with changing temperature, the rate of desorption may be dependent on the combined effect of both factors. Removal of desorbing gas can be speeded up by employing fans and blowers to force fresh air through the material. Natural ventilation may be hastened by taking the goods out of doors where advantage can be taken of wind, thermal air currents and the warming effect of sunlight. Some of the residual fumigant, usually small in quantity, may not be desorbed because of chemical reaction with the material.

Chemical Reaction

If chemical reaction takes place between the gas and the material, new compounds are formed. This reaction is usually characterised by specificity and irreversibility. If the reaction is irreversible, permanent residues are formed. Examples are the reaction between hydrogen cyanide (HCN) and the reducing sugars in dried fruits with the formation of cyanohydrins or the appearance of inorganic bromide compounds after treatment of some foodstuffs with methyl bromide. Because this type of reaction is essentially chemical it may be expected that its intensity varies directly with the temperature. This assumption has been confirmed by observation. Dumas has reported proportionately less fixed bromide residues in fruits as the temperature of fumigation was reduced from 25 to 4°C.

Lindgren et al found an increase in the bromide content of wheat as the temperature during fumigation rose from 10 to 32°C. In recent years attention has been focused on the nature and possible effects on human beings of insecticidal residues appearing in foodstuffs. World-wide interest in this problem is reflected in the fact that international organisations such as the Food and Agriculture Organisation of the United Nations (FAO) and the World Health Organisation (WHO) have set up special committees to investigate and report on the nature and significance of residues formed in foodstuffs as the result of the application of pesticides at different stages prior to human consumption. These special committees review a number of pertinent factors involved in the use of each pesticide.

Important factors, among others, are the toxicological significance of any residues formed and the average fraction of the total diet likely to be constituted by a food containing this residue. Through their Codex Alimentarius Committee these organisations undertake "to recommend international tolerances for pesticide residues in specific food products." Such recommendations are not binding on Member Nations of these organisations but are intended to be used as guides when particular countries are formulating their own regulations for pesticide residue tolerances.

Other Effects on Materials

Apart from the question of significant residues in foodstuffs, there is the problem of other effects which have a direct bearing either on the choice of the particular fumigant or on the decision as to whether fumigation is possible at all. The main types of reaction are:

Physiological Effects

1. Nursery Stock and Living Plants
   (a) Stimulation of growth
   (b) Retardation of growth
   (c) Temporary injury and subsequent recovery
   (d) Permanent injury, usually folalowed by death
2. Seeds
   (a) Stimulation of germination
   (b) Impairment or total loss of germination
   (c) Poor growth of seedlings from germinated seeds
3. Fruit and Venetables
   (a) Visible lesions
   (b) Internal injury
   (c) Shortening of storage life
   (d) Delay of ripening
   (e) Stimulation of storage disorders
4. Infesting Organisms
   (a) Death
   (b) Stimulation of growth or metamorphosis
   (c) Delay in development
   (d) Stimulation of symptoms of disease (so-called "diagnostic effect")

Effects on nonliving materials

1. Production of foul or unpleasant odours in furnishings or materials stored in premises.

2. Chemical effects that spoil certain products (for example, some fumigants render photographic films and papers unusable).
3. Reaction with lubricants followed by stoppage of machinery (clocks will often stop after fumigation with HCN).
4. Corrosive effects on metals (phosphine reacts with copper, particularly in humid conditions).

Difference between Dosages and Concentrations

There should be a clear understanding of the difference between dosage and concentration. The dosage is the amount of fumigant applied and is usually expressed as weight of the chemical per volume of space treated. In grain treatments, liquid-type fumigants are often used and the dosage may be expressed as volume of liquid (litres or gallons) to a given volume (amount of grain given as litres or bushels) or sometimes to a given weight (quintals, metric tonnes or tons). From the moment that a given dosage enters the structure being fumigated, molecules of gas are progressively lost from the free space either by the process of sorption and solution described above or by actual leakage from the system, if this occurs.

The concentration is the actual amount of fumigant present in the air space in any selected part of the fumigation system at any given time. The concentration is usually determined by taking samples from required points and analysing them. It may thus be said that the dosage is always known because it is a pre-determined quantity. Concentration has to be determined because it varies in time and position according to the many modifying factors encountered in fumigation work. Three methods of expressing gas concentrations in air are in common use: weight per volume, parts by volume and percent by volume.

Weight and Volume

For practical designation of dosages, this is the most convenient method because both factors—the weight of the fumigant and the volume of the space—can be easily determined. In countries using the metric system, this is usually expressed in grammes per cubic metre (g/m^3), whereas in countries using the British system of weights and measures, expression is usually in terms of pounds or ounces avoirdupois (avdp) per 1 000 cubic feet (lb/1 000 ft^3 or oz/1 000ft). By a fortunate coincidence in units of measurement, grammes per cubic metro are, for all practical purposes, equal to ounces per thousand cubic feet. Thus, recommended dosages can readily be converted from one system to the other.

Parts by volume and percent by volume will be discussed together because both modes of expression give the relative numbers of molecules of gas present in a given volume of air. The values for both modes have the same digits, but the decimal points are in different places. Parts per million of gases in air are used in human and mammalian toxicology and in applied industrial hygiene. Percent by volume is used in expressing the flammability and explosive limits of gases in air.

Toxicity of Fumigants

As far as is known at present, fumigants enter the insect mainly by way of the respiratory system. The entrance to this system in larvae, pupae and adults is through the spiracles, which are situated on the lateral surfaces of the body. The opening and closing of the spiracles are under muscular control. To enter insect eggs, gases diffuse through the shell of the egg or through specialised "respiratory channels". It has been shown that some gases may diffuse through the integument of insects, but at present the comparative importance of this route for the entry of fumigants is not known. It is known that the poisoning of an insect by a fumigant is influenced by the rate of respiration of that insect; any factor that increases the rate of respiration tends to make the insect more susceptible.

Environmental Impacts on Fumigants

Impacts of temperature

The most important environmental factor influencing the action of fumigants on insects is temperature. In the range of normal fumigating temperatures from 10 to 35°C, the concentration of a fumigant required to kill a given stage of an insect species decreases with the rise in temperature. From the purely biological standpoint, this is mainly due to the increased rate of respiration of the insects in response to the rise in temperature. Also, as pointed out previously, physical sorption of the fumigant by the material containing the insects is reduced and proportionately more fumigant is available to attack the insects.

Therefore, within the range mentioned, conditions for successful fumigation improve as the temperature rises. At temperatures below 10°C, the situation is more complicated. Below this point, increased sorption of the gas by the body of the insect may counterbalance the effects of decrease in respiration, and also the resistance of insects may be weakened by the effects of exposure to low temperatures. With some fumigants, less gas is required to kill certain species as the temperature is raised or lowered on either side of some point at which the insects are most tolerant.

However, with others, toxicity to the insects declines as the temperature falls; for example, with methyl bromide there is a moderate decrease in toxicity down to the boiling point and below this temperature effectiveness drops off sharply so that the amount of gas required to kill the insects increases dramatically. At lower temperatures sorption of the fumigant by the infested material is increased and more fumigant must be applied to compensate for this. Also, diffusion of a gas is slowed down in relation to reduction in temperature.

It is important to bear in mind that the results of a fumigation may be influenced not only by the temperature prevailing during the treatment, but also by the temperatures at which the insects are kept before and after treatment. If the insects have been kept in a cool environment, their metabolic rate will be low. If they are immediately fumigated at a higher temperature, their physiological activity may still be influenced by their previous history, and the uptake of the poison may not be as great as if they had been kept at the temperature of fumigation for a long time previous to treatment. These phenomena can be of practical significance, particularly for certain species of insects that may go into a state known as diapause.

For insects in this state, tolerance to some fumigants, e.g. methyl bromide and phosphine, may be several times greater than for non-diapausing insects. For other species not in diapause, toxicity is usually found to be closely dependent on the temperature of the fumigation. A fumigator must have some knowledge of the previous history of the infested material as well as the species to be treated if he or she is to apply the recommended fumigation treatments most effectively. In all treatments, the material should be warmed to the treatment temperature for several hours to bring the insects to corresponding physiological activity before fumigating.

If species disposed to the state of diapause are present the dosage and exposure applied should be increased to a level that will kill the most tolerant insects. Under experimental conditions, variations in postfumigation temperatures have been observed to influence insect mortalities, but the effects are more complex than those observed in the study of prefumigation temperatures.

Impacts of humidity

From the present knowledge of insect toxicology, it is not possible to make any general statements about the influence of humidity on the susceptibility of insects to fumigants. Variations in response at certain humidities have been observed not only between different species subjected to different fumigants but also between stages of the same species exposed to a single fumigant. However, variations due to humidity are not so important in practice as those due to temperature.

Impacts of carbon dioxide

Carbon dioxide, in certain concentrations, may stimulate the respiratory movements and opening of spiracles in insects. A number of authors have shown that addition of carbon dioxide to some of the fumigants may increase or accelerate the toxic effect of the gas. With each fumigant acting on different insects, there seems to be an optimum amount of carbon dioxide needed to provide the best insecticidal results.

Excessive amounts of carbon dioxide tend to exclude oxygen from insects and thus interfere with the action of the fumigants. With certain fumigants, such as ethylene oxide and methyl formate, the addition of carbon dioxide may work to advantage both by reducing the fire or explosion hazards and by increasing the susceptibility of the insects. On the other hand, with fumigants that are nonflammable, the advantages of adding carbon dioxide may be offset by the extra cost and work required to handle the additional weight of containers.

Protective Narcosis

Some fumigants can produce paralysing effects on insects that may alter the toxicity of these or other fumigants. In the use of hydrogen cyanide (HCN) against insects, it has been shown that, if certain species are exposed to sublethal concentrations before the full concentration is applied, the resulting fumigation is less effective than one in which the insects are subjected to the full concentration from the very beginning. A similar protective effect can also occur with the fumigant phosphine if insects are exposed to excessive concentrations during a treatment.

Also, insects that have been narcotised by sublethal concentrations of HCN have been found to be protected from lethal treatments with other fumigants, e.g. methyl bromide and phosphine. This effect has been referred to as "protective stupefaction" or "narcosis". Although phosphine itself can narcotise insects it does not, however, protect them from the action of methyl bromide as does HCN; in fact, phosphine and methyl bromide can be used together as a "mixture" to enhance the effectiveness of each other. From the practical point of view the phenomenon of narcosis is important because it can reduce the effectiveness of certain fumigants. However, steps can be taken to avoid problems of this nature:

1. In fumigations with HCN the maximum concentration attainable from a recommended dosage should be achieved as soon as possible at the beginning of the treatment.
2. HCN should not be applied with other fumigants such as methyl bromide or phosphine, if the maximum toxic effect is to be achieved.
3. Excessive concentrations of phosphine likely to produce a protective narcosis should not be used.

Susceptibility of insects

It has often been observed that there may be fluctuations in the susceptibility of populations of insects to a given poison. Some of the reasons are known, while others need further clarification. Two important factors are undoubtedly seasonal changes in climate and the effect of nutrition. The susceptibility of insects may be greatly influenced by the quality of the food they consume. It also has been observed with some insects that a certain amount of starvation may make them more, rather than less, resistant to fumigants. In practical work it is well to know that fluctuations in resistance may occur. The alert operator must always be on the lookout for any changed conditions that may necessitate modification of recommended treatments.

Comparative toxicity of fumigants

Apart from the influence of the environment, there is a great variation in susceptibility of different species of insects to different fumigants. The successive stages of a given species may also vary greatly in response. A large number of studies have been made under laboratory conditions to determine the relative susceptibility of insects to different fumigants.

Bowley and Bell have reported on the toxicity of twelve fumigants to three species of mites infesting grain. The treatments recommended here are based on laboratory or field trials that have been confirmed, in many instances, by the results of successful application in practice. Note that all recommended treatments refer to specific insects or their stages or, in some cases, to clearly defined groups of insects. There is, therefore, no guarantee of the success of any attempts to apply a treatment outside the limits given in the recommended schedules.

Resistance of Insects

Many species of insect have the ability to develop resistance to certain insecticides. With fumigants this problem of resistance is a matter of increasing concern; in a global survey of stored grain pests, resistance to both of the major fumigants, phosphine and methyl bromide, was found in a number of insect species. Collections of 849 strains of insects from 82 countries showed that 20 percent of the insects had some resistance to phosphine and 5 percent to methyl bromide. The highest level of resistance (10-12 times normal) was found in the lesser grain borer Rhyzopertha dominica (F.).

It was concluded from this survey that resistance to fumigants was, as yet, limited in extent and often at marginal levels, but that it was of considerable significance as it posed a real threat to the future use of fumigants as control agents. Research in laboratories has shown that a number of destructive stored product insects can develop appreciable resistance to fumigants. Selection of the granary weevil (Sitophilus qranarius) has produced a strain with more than 12 fold resistance to methyl bromide.

A strain of the red flour beetle, Tribolium castaneum, developed a 10-fold resistance to phosphine in six generations. There is recent evidence, from field studies in India and Bangladesh, of the development of resistance to phosphine in the Khapra beetle and other stored grain pests. Resistance to fumigants is of concern because of the great value of fumigants for pest control and because of the very limited number of materials available. Even low levels of resistance in species of insects that are cosmopolitan and easily transported to other parts of the world could be of serious consequence. In view of the importance of resistance to fumigation, a brief and simplified account of some features of the problem are given below.

When a population of insects is exposed to an insecticide some individuals are killed more easily than others. The insects that are more difficult to kill may survive after the treatment and produce offspring that are also hard to kill. These insects are said to be more tolerant because they can withstand above-average doses of the poison. If a population is repeatedly treated with the same insecticide and each new generation has increasingly higher tolerance, a "resistant" strain is produced. Resistance is a genetic characteristic that is passed on from one generation to the next.

In the laboratory, resistance is produced by treating a population to kill most of the insects, breeding the tolerant survivors to produce a new generation, re-treating and continuing the process until a resistant strain is obtained. This process is known as selection for resistance. A number of strains of insects with resistance to different fumigants have been produced in this way. In the field, resistance to fumigants can develop in the same way. In a grain bin, on a cargo ship or any other place where a resident population of insects is treated over and over again with the same fumigant, resistance might develop.

Insects that are not killed may produce new generations with increasingly greater tolerance. Generally, resistance does not develop as readily in wild populations as in the laboratory because the selection process.may be irregular and because they may interbreed with nontreated susceptible insects. However, the fact that resistance has been discovered in wild populations indicates the possibility that further resistance may develop where fumigants are used regularly.

Nature of resistance

Resistance is an inborn characteristic that allows individual insects to tolerate above average doses of a poison. Resistant insects usually are similar in appearance and have the same habits as susceptible insects. Normally, they can only be distinguished by their ability to tolerate excessive concentrations of the fumigant. Tests have been designed for detecting and measuring resistance to fumigants. An important feature in resistance is the ability to tolerate the effects of more than one poison.

Insects that have resistance to one fumigant can, in some cases, also be resistant to other fumigants. This characteristic, known as "cross-resistance" is demonstrated by the data in Table 3. It can be seen that granary weevils selected with methyl bromide were also resistant to several other fumigants, and the levels of cross-resistance were all significant in terms of practical control. Such cross-resistance was not found, however, in insects selected with phosphine or ethylene dibromide.

Table 3. Response of Methyl Bromide-Resistant Granary

Fumigant	*Resistant*	*Normal*	*Tolerance ratio*
Methyl bromide	19.7	3.6	5.5
HCN	16.4	8.2	2.0
Acrylonitrile	4.9	1.05	4.7
Ethylene oxide	20.1	4.1	4.8
Chloropicrin	6.6	3.9	1.7
Phosphine	13.0	2.2	5.9
Ethylene dibromide	8.5	2.85	3.0

Resistance testing

For routine monitoring to detect the initial appearance of resistance in wild populations of stored product beetles, it is convenient to use a discriminating dose, which is expected to kill all susceptible specimens. The dose chosen is that corresponding to slightly above the LD (99) 9 obtained from the regression line for susceptible beetles allowing for, in the case of phosphine, what appears to be inherent variability of response.

When using a discriminating test with fumigants it is always advisable to make provision for abnormal concentrations. If a concentration is obtained that is less than the discriminating concentration, this will be revealed by abnormal survival in the susceptible reference strain. Abnormally high concentrations may be revealed by the inclusion in the tests of a reference strain with slightly greater tolerance to the fumigant than the susceptible reference strain on which the discriminating dose is based, approximately x 1.5 for methyl bromide tests and x 2.5 for phosphine tests. An alternative approach is to use three dosages, one at the discriminating dose, one at the approximate LD(90) level and the other at an equivalent level above the discriminating dose.

In regular monitoring for resistance, it should be detectable even when only a small proportion of resistant individuals is present. A minimum of 100 insects in two batches of 50 should be used per sample. Limited numbers of insects may not be sufficient to detect low levels of resistance. Therefore, additional samples should be obtained, if possible. If, however, there is suspicion of serious resistance (e.g. from failure of treatments) a test with small numbers (10 to 20) may provide valuable early indication.

The insects are exposed to the discriminating dose for the appropriate period in the usual way. If all of them are dead at the end of the posttreatment holding period, the sample can be classified as "no resistance detectable", and the medium in which they were held is put into a hot-air oven to destroy the culture. On the other hand, the presence of surviving insects at the end of the test should be regarded as prima facie evidence of resistance and investigated further.

Confirming resistance

The appearance of unaffected insects in a discriminating test could be due to the presence of unusually tolerant individuals from a normal population. Provided that the conditions of exposure, the physiological state of the insects and the dosages are consistent, the probability of a single insect in a batch of 100 being unaffected due to chance is less than 0.1 (e.g. less than once in 10 tests). It is important to determine whether incomplete response is due to such causes or to genuine resistance. This can be checked as follows:

1. The test can be repeated using further samples from the same field population. The chances of adventitious failure to respond by a single individual in each of successive tests decline

progressively. Survival of two or more indviduals throughout is even less likely. Therefore, the continued appearance of a proportion of unaffected individuals can be considered as proof of resistance.

2. Alternatively, the insects which were unaffected in the discriminating test may be kept and used for breeding a further generation. If their reaction is actually due to resistance, it will be found that a substantially larger proportion of their progeny will fail to respond to the discriminating dose.

Table 4. Normal Susceptibility data obtained for methyl bromide and phosphine, with discriminating dosages

	LD(50)	LD(99.9)	Discriminating dosage
Methyl Bromide		*mg/l*	
(Exposure period - 5 hours)			
Sitophilus oryzae (L.)	3.6	4.8	6
S. zeamais Motsch.	3.2	5.4	6
S. qranarius (L.)	5.1	7.5	9
Rhyzopertha dominica (F.)	4.0	7.4	7
Tribolium castaneum (Herbst)	8.4	11.7	12
T. confusum Duv.	8.6	11.2	13
Oryzaephilus surinamensis (L.)	5.8	8.5	9
O. mercator (Fauv.)	5.8	8.5	9
Phosphine			
(Exposure period - 20 hours)			
Sitophilus oryzae	0.011	0.039	0.04
5. zeamais	0.007	0.013	0.04
5. qranarius	0.013	0.041	0.07
Rhyzopertha dominica	0.008	0.028	0.03
Tribolium castaneum	0.009	0.028	0.04

T. confusum	0.011	0.029	0.05
Oryzaephilus surinamensis	0.012	0.036	0.05
O. mercator	0.011	0.034	0.05

When these tests indicate that a population of insects is resistant, then extensive testing should be carried out to determine the degree of resistance present.

Different Ways to Avoid Resistance

Precautions can be taken to reduce the possibility of insects developing resistance to fumigants. Perhaps the most effective measure involves alternate control practices that do not require chemicals. Good sanitation procedures, proper storage conditions, insect resistant packaging and all other measures that prevent infestations from developing can do much to reduce the need for fumigants.

Where fumigants have to be used on a regular basis, close guard should be kept against control failures. Complete control of all insects in a treatment is the best insurance against resistance. Periodic checks for resistance should be made in areas that are fumigated regularly. If signs of resistance begin to appear then every effort should be made to eradicate the population. The measures necessary for eradication will vary in different situations; they may involve a number of procedures using both chemical and non-chemical means.

Rotation of fumigants may be effective in some instances, especially if crossresistance is not a problem. For example, methyl bromide might be used at intervals in a control programme that relies mainly on phosphine. One measure that is not advisable in dealing with resistance problems involves increased dosing. Such practices as doubling the dose of fumigant to achieve an economic level of control can magnify the problem unless complete eradication is assured. Any insects surviving increased doses may develop even higher levels of resistance than would occur with the normally recommended tareatment.

Fumigation and Pest Management Programmes

The use of certain fumigants has been restricted in some countries because of suspected adverse effects. Excessive use of fumigants or the misuse of them to cause accidents and produce adverse publicity is likely to bring about even greater restrictions in their use.

Fumigation is just one of a number of methods that can be used for controlling pests in stored products. The best control is likely to be obtained when all appropriate measures are taken to eliminate pest organisms. In an effective pest management programme, methods of prevention and control are integrated to give maximum protection of goods at the lowest possible cost. Other ways that have been found effective in preventing and controlling infestations are as follows:

1. Sanitation.
2. Exclusion of pests.
3. Low temperature - "freeze-outs", refrigeration, aeration.
4. High temperature - heating of mills.
5. Moisture control - grain drying.
6. Aeration - cooling, drying, elimination of temperature gradients.
7. Protectants - chemicals, inert dusts, natural compounds.
8. Residual or contact insecticides.

9. Atmospheric gases - carbon dioxide, nitrogen.
10. Gamma radiation, radio and sonic waves, microwaves, infra-red radiation.
11. Pheromones.
12. Insect growth regulators.
13. Insect pathogens.
14. Predators.
15. Insect resistant packaging.
16. Resistant varieties.

An effective integrated pest management system should begin with comprehensive planning to include all aspects of the problem, followed by the application of preventive and control methods. For example, the planning of pest management for a commodity like farm stored grain may be divided into five major categories:

- exclusion of the pest organism;
- inspection procedures;
- good housekeeping and sanitation;
- physical and mechanical control;
- chemical control

Infestation problems can often be reduced by careful planning so that the possibilities of pest organisms reaching the commodity will be kept to a minimum. Location of the storage relative to sources of infestation is important, as well as quality of the structure. Well-built storages, with a minimum of sites where debris can accumulate and insects develop, are desirable. Other features of the storage that should be considered include facilities for conditioning such as aeration systems or driers, provision for proper inspection and cleaning and appropriate facilities for pest control procedures.

Preventive and control methods may include the following:

1. Use of sound structures for storage of commodities.
2. Maintaining clean conditions around storages.
3. Removal of residues of grain or other material from the storage facility four to six weeks prior to storing newly harvested produce.
4. Spraying of storage with approved residual insecticide after removal of food residues.
5. Storage of commodity in a condition suitable for optimum storage, e.g. grain is best stored at low moisture levels.
6. Treatment with appropriate insecticide protect ant at the time of storage may be desirable.
7. Use of aeration or other procedures to cool grain and maintain uniform temperature below those favourable for insect development.
8. Regular inspection to determine
   (a) evidence of insect activity
   (b) accumulation of moisture
   (c) changes in temperature
9. If insects are detected, grain should be fumigated; where field infestation occurs grain should be fumigated within six weeks after harvest.

By careful planning and management, fumigation may be incorporated into food preservation systems so that fumigants can be used more effectively and safely than when used independently. They should never be used as a substitute for sound management and good sanitation procedures. The benefits derived can include reduced cost of storage with improved food quality, reduced residues in food materials, greater occupational safety and less environmental contamination.

All of these benefits are of great concern to the general public and will be factors that have to be taken into consideration in the future use of fumigants. The ultimate goal in the control of pests in stored products should be to so improve the methods of handling, storing and processing commodities, that the need for pesticides will decrease. Fumigants will then only be needed when unavoidable infestations are encountered.

2

Chemicals Used as Fumigant

Fumigation makes use of some of the most toxic chemicals known to modern civilisation. Fumigants are toxic to all forms of life; a handling error on your part can kill you or the person working with you. Consequently, the Environmental Protection Agency (EPA) drastically changed the way fumigants are to be handled. Some products such as ethylene dibromide (EDB) and carbon disulfide are no longer registered for use and the remaining fumigants have been classified as restricted use pesticides.

A minimum of two trained people must be present if entry of a structure is necessary during fumigation or if you must reenter a partially aerated structure. Approved respiratory protection is required when working in areas where the fumigant concentration is at or above the threshold limit value (TLV), or when the fumigant concentration is unknown. Monitoring equipment is needed to determine fumigant concentrations prior to entry into a structure under fumigation and before warning signs are removed.

Methyl Bromide

The insecticidal value of methyl bromide was first

reported by Le Goupil in France. During the 1930s it was widely adopted for plant quarantine purposes because many plants, vegetables and some fruits were found to be tolerant to concentrations effective against the insects concerned. More recently it has been used extensively as an industrial fumigant for stored products, mills, warehouses, ships and railway cars. For this purpose it has now largely replaced hydrogen cyanide.

Methyl bromide has also been used as a sterilising agent, although it has approximately one tenth the activity of ethylene oxide against bacteria and fungi. Its use for the sterilisation of space vehicles in combination with ethylene oxide has been reported by Vashkov and Prishchep. At concentration x time products considerably higher than those needed to kill insects, methyl bromide may also control microorganisms such as *Aspernillus* spp. and *Penicillium* spp. in foodstuffs. Methyl bromide is not as toxic to most insect species as are some other commonly used fumigants, such as Hydrogen cyanide (HCN), acrylonitrile and ethylene dibromide. Nevertheless, other properties make methyl bromide an effective and versatile fumigant.

The most important of these is its ability to penetrate quickly and deeply into sorptive materials at normal atmospheric pressure. Also, at the end of a treatment, the vapours dissipate rapidly and make possible the safe handling of bulk commodities. Another important property is the fact that many living plants are tolerant to this gas in insecticidal treatments. Methyl bromide is nonflammable and nonexplosive under ordinary circumstances and may be used without special precautions against fire. Because methyl bromide has a comparatively low boiling point and is not greatly sorbed by many materials, it may be used for low temperature treatments that are not practicable with many other fumigants.

Properties

Treatments of a wide range of commodities may be conducted at temperatures down to 4°C, or even lower in some instances. At normal fumigation concentrations methyl bromide is odourless. This disadvantage is sometimes overcome by mixing it at the time of packaging with a warning gas such as chloropicrin. The chloropicrin usually constitutes 2 percent of the mixture.

The effect of methyl bromide on humans and other mammals appears to vary according to the intensity of exposure. At concentrations not immediately fatal, this chemical produces neurological symptoms. High concentrations may bring about death through pulmonary injury and associated circulatory failure. The onset of toxic symptoms is delayed, and the latent period may vary between 0.5 to 48 hours, according to the intensity of the exposure and the personal reaction of the patient.

Contact of the human skin with the liquid or strong concentrations of the gas may cause severe local blistering. Against insects, methyl bromide appears to exert its principal toxic effect on the nervous system. As in humans, the onset of poisoning symptoms may be delayed, and with many species of insects definite conclusions as to the success of the treatment should be delayed for at least 24 hours. Methyl bromide is also effective against mites (Acarina). In the treatments in which living plants and flower bulbs are tolerant, the eggs of mites may be resistant and repetition of fumigation may be necessary.

In laboratory experiments with an intense electric spark, it was found that methyl bromide had a range of flammability from 13.5 to 14.5 percent by volume in air. This range has been widely quoted in scientific and trade literature, and the impression has been created that methyl bromide may be flammable or explosive under certain conditions in air. However, in the same series of tests it was found that mixtures of this gas with air in any proportions are nonflammable when ignition is attempted with a flame. Methyl bromide has proved successful as a fire extinguisher.

It should be pointed out that, in the absence of oxygen, liquid methyl bromide reacts with aluminium to form methyl aluminium bromide. This compound ignites spontaneously in the presence of oxygen. The explosion of a large tank containing methyl bromide was attributed to the reaction of this compound with the aluminium stem of a measuring instrument beneath the surface of the liquid. Methyl bromide should never be stored in cylinders containing any appreciable amount of the metal aluminium and aluminium tubing should not be used for application of the fumigant.

Effect on Plant

Methyl bromide is one of the few fumigants that may be used safely on a wide range of living plants without producing harmful effects. However, there is a limited number of genera, species or varieties of plants susceptible to injury. Methyl bromide has been employed as a seed fumigant because of its ability to penetrate into large consignments of sacks and bags. Under some circumstances, however, treatment with this fumigant has resulted in loss of viability. Also, germination may be delayed or the vitality of young plants impaired. Intensive investigations during recent years have disclosed that injury to germination and retardation of growth of seedlings are directly related to abnormally high temperature, dosage of fumigant, length of exposure and moisture and oil contents of the seed.

Blackith and Lubatti laid down a useful general maxim for methyl bromide: "If the seed is dry enough to store, it is dry enough to fumigate." They kept cereal seeds which had been fumigated at 8, 11, 14, and 18 percent moisture content for six years. The seeds were stored at 20°C and at the same moisture content at which they were fumigated. Germination tests were made after 6 months, 3 years and 6 years. For the most part good survival was recorded for both fumigated and nonfumigated seeds at the two lower moisture contents. Rye retained its viability for 3 years when stored dry, but both treated and untreated rye showed almost complete loss of -termination after 6 years.

Powell tested the tolerance of 40 varieties of vegetable, cereal, fodder and grass seeds with methyl bromide and he concluded that, although the lower values of dosage, temperature and moisture resulted in better germination, most seeds could be safely fumigated at moisture contents below 13 percent and in many cases up to 16 percent. Treatment of vegetable seeds with mixtures of methyl bromide and carbon dioxide and under partial vacuum likewise had no detectable effect on germination. Repetition of fumigation on a given lot of seed should be avoided, if possible; it has been shown that more than one fumigation may cause reduction in percentages of germination.

Repeated fumigation can also have adverse affects on the subsequent growth and yield of plants grown from the treated seeds. In tests on white and yellow maize, Joubert and Du Toit reported that all treatments caused significant reduction in yields from white maize and that, while the apparently hardier yellow maize showed little effect at time of germination, there was a significant reduction in yield after the second and third fumigation. If repeat fumigation is considered, it is suggested that germination and possibly growth tests be made to determine if injury has occurred during the first treatment. Joubert and Du Toit advocate extreme caution with the application of fumigants to seed and they further suggest that a contact insecticide having no adverse effect on the seed should be considered instead of fumigation.

Methyl bromide may be used to fumigate some growing crops to control pest organisms. The development of light-weight plastic sheets has made it possible to cover sizable areas of land so that the fumigant can be contained for sufficient time to effect a treatment. Large beds of strawberry plants infested with cyclamen mite have been successfully treated in this way and methods for control of European pine shoot moth on ornamental pines have been described. It is estimated that nearly 95 percent of the nursery stock and other plants being moved in commerce are tolerant to dosages of methyl bromide that will kill the insects or other pests involved.

In a few genera of plants, all species may be intolerant while in others only a few species or varieties are known to be affected. Latta and Cowgill tested 441 species of glasshouse plants with methyl bromide and found that 414 (93.9 percent) were not injured and 27 (6.1 percent) sustained varying degrees of damage; of those, five species were severely burned. In tests on improved citrus stock, Riviello and Rhode found that exposure to 16 mg/l for 2 hours at 25 - 30°C and 90 percent RH for control of citrus black fly caused little damage to tender growth..

There are several factors that influence the unfavourable response of growing plants to methyl bromide; if they are avoided, even fairly susceptible varieties will be tolerant or, at worst, only lose foliage that will be replaced by new growth. The most important adverse factors are as follows:

1. Low humidity during treatment. Relative humidity in the chamber should be held above 75 percent during fumigation.
2. Dry soil around roots. Less damage is sustained if the soil is moist, mainly because the roots are less susceptible to injury. If there are no insects to be controlled in the soil, it may be completely drenched. Waterlogged soil will not permit the penetration of methyl bromide, and sometimes it is desirable to protect the roots in this way. On the other hand, moist soil allows the rapid penetration of the fumigant.
3. Excessive air currents during fumigation, or during the post-treatment aeration period, aggravate injury. It is recommended that circulating and ventilating fans or blowers be operated for the minimum length of time required for distributing the fumigant evenly or for removing toxic concentrations after treatment.
4. Some species of plants may be fumigated only when fully dormant; others are tolerant at any normal stage of growth.

Plants in active growth are more subject to injury than are dormant plants. As long as plants are fully dormant, there is less danger of damage from methyl bromide, but at the time of breaking dormancy, there may be a period of susceptibility to injury. Coniferous evergreens are particularly liable to sustain severe damage at this critical stage. Great care must be exercised in the choice of the time for nursery stock treatments. It should be borne in mind that while the failure of subsequent growth of treated plants or any other injury may be attributed to the fumigation treatment, it may also be caused by handling or shipping. Claims that injury to nursery stock is due to the fumigation process should be investigated closely to ascertain whether some other factor or combination of factors may not be responsible.

Methyl bromide may be used to fumigate narcissus and other bulbs to control insects such as the greater and lesser bulb flies. The treatments effective against these pests do not injure the bulbs. This fumigant is also effective against bulb mites. However, the eggs are resistant to methyl bromide and treatments which are strong enough to kill the eggs are completely destructive to the bulbs. For complete control, the regular treatments may be repeated after 10 to 14 days or when the eggs have hatched. The repetition of the mild treatment causes no injury to the bulbs. Purnell and Hague reported that fumigations with methyl bromide effective against the stem nematode are injurious to narcissus bulbs.

Effect on Fruit

Methyl bromide has been used widely for quarantine fumigations of fresh fruit. It has been found that some fruits, or certain varieties, are susceptible to injury. The differences in varietal susceptibility are particularly noticeable in apples. External damage often takes the form of brown lesions or small round spots on the skin caused by increased or decreased pigmentation in the normal colour of the fruit. Internal injury appears usually as browning of the tissue. With some fruit, ripening and pigmentation are stimulated by low dosages and retarded by high ones.

Fruit may vary in susceptibility from one season to another; this is believed to be due to variations in the physiological condition of the fruit. Some workers have been able to detect slight off-flavours or lessening of flavours in fruit subjected to methyl bromide, but it is unlikely that these would be easily detected by the public to the extent that marketing would be affected. The tolerance of deciduous fruit to methyl bromide and other fumigants is discussed fully by Claypool and Vines. While some fruits are susceptible to injury by methyl bromide, others are tolerant and are treated with this fumigant in quarantine procedures.

Methyl bromide is particularly useful when treatments at low temperatures are required. Details of treatments for tolerant fruit are given in Schedule H. It is recommended that fumigation of fruit with methyl bromide should not be carried out on a commercial scale until careful preliminary experiments have indicated that the particular kinds or varieties concerned are tolerant under the full range of conditions likely to be encountered in practice. At dosages considerably lower than those needed to kill insects, methyl bromide has been successfully used for killing rats and mice in fruit storage units. With the treatments given in Schedule T (Rodents), injury to fruit is not likely to occur.

Effect on Vegetables

Both fresh and dry vegetables are generally tolerant to insecticidal treatments with methyl bromide. This fumigant has been used extensively for the treatment of both early and late varieties of potatoes to control the larvae of the tuber moth Phthorimaea operculella. In treatments with concentration x time (c x t) products of 85 mg in/l, which are completely effective against these insects, tubers of all varieties tested at all stages of maturity have proved to be tolerant.

Adults and pupae of the Colorado potato beetle can also be controlled without injury to the potatoes. Sweet potatoes were found by Phillips and Easter to be sensitive to insecticidal treatments under some conditions. It was noted that newly harvested sweet potatoes were likely to be severely injured, whereas those cured for 15 days or more or kiln dried at 27 to 30°C, were not always injured. Breakdown of cured sweet potatoes after fumigation was associated with postfumigation storage temperatures. Material held at 21 to 27°C was affected less than that moved to ordinary storage at lower temperatures. Proper ventilation after fumigation decreased the amount of the breakdown.

Effect on Cereals

Methyl bromide is widely used for the fumigation of almost every type of cereal and cereal product. Because it penetrates densely packed materials, it is especially useful for the treatment of flours and meals. Although methyl bromide will react with the protein fraction of wheat, a number of studies have shown that a single treatment at recommended dosages has little or no adverse effect on food value or bread making quality. Even repeated fumigation of wheat stored over a 3 year period with a total of 8 treatments caused no significant effect on the vitamin B-6 components or Tocopherols.

However, there was a gradual increase in inorganic bromide residue and there were some indications of minor changes in physical qualities of bread made from this wheat. Some members of a taste panel detected a stale aroma in laboratory-type breads and crumbly texture in rolls made from wheats exposed to repeated methyl bromide fumigations. It has been noted that bread made from flour fumigated with excessive dosages of methyl bromide may have a foreign odour, and if the bread is toasted an unpleasant off-flavour may be produced. Occurrence of this phenomenon is rare and sporadic; Brown et al., reviewed a number of reports of this taint encountered under commercial conditions and carried out a number of tests in the laboratory and they came to the following conclusion:

"It would now appear that some taint is possible even at dosages normally used commercially for insect control. This tainting usually takes the form of abnormal odours when the hot loaves are removed from the oven. The taste of the bread when cold is quite unaffected at these levels of treatment but a faint odour may still be detected if the bread has been wrapped in polythene or similar material. The avoidance of taint is made more difficult by the likelihood of uneven distribution of fumigant in many types of treatment. However, the risks become very slight if care is taken to limit the level of treatment, if restrictions are placed on repeated fumigation, and if appropriate dilution of fumigated flour by unfumigated flour is arranged where necessary. Over many years very large tonnages of flour have been fumigated with methyl bromide without difficulty and the number of complaints that have come to light is extremely small. These might have been avoided if the precautions suggested had been followed."

As general precautions in the fumigation of flour with methyl bromide great care must be taken not to exceed recommended concentrations or treatment periods; to ensure that even distribution of the fumigant be effected as quickly as possible after the beginning of the treatment and that aeration of the entire stock be conducted quickly and thoroughly immediately on termination of the treatment. High moisture content of the flour may also be an important factor, as was suggested by the work of Hermitte and Shellenberger who, however, used excessive doses of methyl bromide greatly above those used in commercial practice.

Effect on Nuts

These are treated regularly with methyl bromide, because they are often best suited for bulk treatment. An interesting and spectacular practice is the treatment of large pyramids of groundnuts awaiting shipment at collecting points in west Africa. Although neither high residues nor taint are produced by normal treatments, excessive dosages or repetitions of treatment should be avoided.

Srinath and Ramchandani reported some off-flavour in walnuts given a second treatment with methyl bromide and more pronounced effects after a third treatment. Shelled nuts with a high oil content, such as cashews, must be treated with particular care. When nuts are stored for prolonged periods in warm weather, fumigation may be repeated. If it is necessary to apply more than two treatments with methyl bromide, a preliminary fumigation of small samples should be made to determine effects on quality. The problem of treatment repetition may be encountered with importations. Almost invariably, nuts and shelled nuts are fumigated in the country of origin before export, often with methyl bromide. If more than one fumigation is required after importation, there may be danger of taint and a trial treatment should be made.

Effect on Dried Fruit

Methyl bromide has been used extensively in recent years for the atmospheric or vacuum fumigation of dried fruit of all kinds. It is well suited for this purpose because it penetrates densely packed materials and diffuses away rapidly after treatment.

Determinations of Methyl Bromide

Rough determinations of methyl bromide concentrations in air, for the purpose of detecting serious leaks or for protecting personnel during fumigation and subsequent aeration, may be made with a device commonly referred to as a halide leak detector or halide lamp. These lamps are easily obtainable, usually from refrigeration supply dealers, because they are also used for detecting leaks of the commonly used Freon-type refrigerants. All these lamps work on the same principle, that a flame in contact with a clean piece of copper will burn with a green to blue flame if vapour of an organic halide is present in the surrounding air.

At increasing concentrations of the halide gas, the colour changes from green to greenish-blue or blue. On the basis of this principle it is a fairly simple matter to improvise lamps, given a clean copper wire and a source of flame. Any improvised lamps must, however, he subjected to careful preliminary calibrations. Various types of detectors are marketed and the difference in construction depends on the fuel they use.

Fuels used are, among others, paraffin (kerosene), wood alcohol, acetylene and propane. A lamp using a disposable or a rechargeable canister of propane gas is the most convenient, although it is not as sensitive as those burning paraffin or alchohol. Propane canisters are light, readily available, contain enough gas for about 10 hours' burning and the lamp does not require priming before it is lit.

None of the lamps is to be relied upon for accurate quantitative determinations. They are useful for indicating immediately dangerous concentrations, for checking the effectiveness of aeration after many types of fumigation and for finding leaks during treatment. They should not be used for regular routine checks on the threshold limits for continuous daily exposure to methyl bromide; for this purpose some method of chemical analysis or an accurate recording device should be used.

Several considerations are important for the proper and reliable operation of lamps.

1. The copper ring must be kept clean, otherwise a green flame may show in the absence of methyl bromide.
2. The flame must burn through the ring, not around it; if the ring is clean, and the air is free of fumigant or dust, the flame should be invisible above the ring.
3. After some time the copper ring deteriorates and must be replaced by a new one.
4. These lamps are not specific for methyl bromide; they react to any organic halogen gas, such as the Freon refrigerants, whose presence in the air may result in erroneous lamp readings. It is advisable to test atmospheres before the fumigant is applied to ensure that no interfering substance is present.
5. Obviously, the lamps cannot be used in the presence of flammable or explosive gases, such as gasoline vapours, or with some organic halide gases, such as methyl chloride and ethylene dichloride, which are also flammable and may be used as fumigants.

6. Lamps should not be used in dust-laden atmospheres. If gas analysis is required from such areas, a sample of the atmosphere may be obtained by opening a polythene bag at the point to be checked and removing the sample to a safe location. Alternatively, a sample may be drawn from the treated area through plastic or copper tubing to a suitable outside area.
7. Colour blind persons cannot detect colour changes in a halide detector flame.
8. If used at night the gas detector flame will have a bluish cast; hence, appropriate allowance for variations in different light conditions should be made.

Gas detector tubes: Where a more precise measurement of the concentration in the region of the threshold limit value is required, a simple method is the use of a gas detector tube. Tubes for methyl bromide are available from several manufacturers in ranges from 2 - 200 ppm and they should be used strictly in accordance with instructions. A fresh tube is required for each determination.

Infra-red analyser: This detector will measure concentrations of methyl bromide well below threshold limit values. It is portable, battery powered, direct reading and has no flame to cause hazards in dust laden atmospheres. This instrument can be used to determine if buildings or other spaces are safe for occupancy.

Gas chrometopraphy: The portable gas chromatograph will analyse methyl bromide at working levels used in fumigation and at low levels down to 0.01 ppm and below.

Thermal conductivity analyser: Instruments of this type are particularly suitable for use with methyl bromide and are, in practice, used mainly with this fumigant. However, they are not sensitive enough to determine the health hazards presented by low concentrations of methyl bromide and are not authorised for this purpose in fumigation codes of safety.

Chemical analysis: A convenient method of chemical analysis for field use is the Volhard titration after the gas has been absorbed in monoethanolamine. Chemically pure monoethanolamine, in the amount of 2 ml for each 0.45 kg (1 lb) of methyl bromide per 28 m′ (1 000 ft^3) in the fumigation space, is placed in a glass bottle fitted with a stopcock. The glass bottle must be able to withstand evacuation; round 1-litre (quart) bottles have been used successfully.

The stopcock is connected to copper or polyethylene tubing that leads from the desired sampling points in the fumigation system. Before a sample is taken, the bottle is evacuated to a pressure of 1 to 2 cm. The flask should stand for at least 2 hours before the titration is started. The bromide is estimated after it has been precipitated by excess standard silver nitrate solution; the excess silver is titrated by standard potassium thiocyanate, ferric alum being used as an indicator.

Laboratory analysis: The Volhard titration method given above is suitable for laboratory determination of vapours. Dumas and Latimer described a method for analysing the vapours by a technique of coulometric titration whereby amounts as small as 17 μg of methyl bromide in a sample may be determined accurately. Berck listed methyl bromide among the fumigants which may be determined by gas chromatography.

Mapes and Shrader gave a chemical method for determining total and inorganic bromide residues in fumigated foodstuffs. The basis of this method is that the bromides are hydrolysed with ethanolamine and the alcohol is removed by evaporation. The residue is ashed with sodium hydroxide and sodium peroxide. The bromide is leached from the ash and determined by the KolthoffYutzy-van der Meulen method. Total bromide ion may also be determined by xray fluorescence or by neutron activation analysis. Selective determination of bromide ion may be carried out with a selective ion electrode or by using gas chromatography. Determination of free methyl bromide in commodities can be carried out by various methods using gas chromatography.

Application of Methyl bromide

Methyl bromide is supplied by manufacturers in the following types of containers:

Steel cylinders with capacities from 5 to 1 800 lb (2.25 to 816 kg). For ordinary field purposes, the 50- or 100-lb (23- or 45-kg) sizes are the most convenient. The cylinders have siphons and at normal temperatures the natural vapour pressure of the fumigant is sufficient to discharge it from the cylinder. Some manufacturers inject an inert gas, such as nitrogen, under pressure into the space above the liquid before the cylinder is shipped; this helps to force the fumigant through lines of piping.

Other makers supply additional valves on the heads of the cylinders so that more pressure from compressed nitrogen or air may be applied at the time of discharge. To apply the desired quantity of fumigant from a cylinder, a set of scales is used to determine the amount being discharged. Spray nozzles, as described by the U.K. Ministry of Agriculture, Fisheries and Food or a sprinkler as outlined by Calderon and Carmi have been found useful for dispersing the fumigant in large spaces.

Cans containing 1 or 1.5 lb (0.45 or 0.68 kg) of the fumigant. Some manufacturers make 0.5 lb cans available on request. A special device for discharging the fumigant from them under natural pressure is obtainable from the supplier and copper or plastic tubing may be attached to carry the methyl bromide to the space being fumigated, as required. This method of application is useful for small-scale operations because the cans are easy to handle and the dosages are easy to compute in terms of the number of cans required.

An opened can must be completely emptied as it cannot be resealed. If the fumigant is discharged as a gas from the can, considerable cooling of the liquid takes place, especially in cool weather, and the temperature may fall below the boiling point of methyl bromide (3.6°C). If the can is immersed in a pail of hot water (not above 77°C), the fumigant is discharged more evenly and rapidly.

Glass ampoules, usually containing 20 ml of methyl bromide, are supplied by some manufacturers. These are useful for small chambers. A plunger or similar device is needed for breaking these ampoules so that the fumigant can disperse in the chamber.

Freezing of the tubes and piping carrying the fumigant sometimes occurs due to the loss of heat from evaporation. This is more likely to occur when the fumigant is carried some distance from the container. In many applications, therefore, the fumigant is led through a heat exchanger after it leaves the container. For small dosages, a suitable exchanger is a hot water bath or pail in which is coiled 15 m (50 ft) of copper tubing. For large-scale operations, a greater transfer of heat is required. Several devices have been described in detail by Hammer and Amstutz.

The essential feature of these heaters is a vaporising chamber of sufficient capacity to convert the liquid methyl bromide into ves as it passes through. At temperatures below 15°C or when large quantities of methyl bromide are used a vaporiser, a 15 m coil of copper tubing, 10 - 15 mm in diameter, immersed in a container of water heated to 65°C may be required. The fumigant should be passed through the tubing at a rate of 1 - 2 kg per minute and liberated as a vapour into the air stream of a fan or blower. Any convenient and safe source of heat may be utilised for these heaters.

Bottled propane and live steam have proved successful. Heat from a flame or electrical heater should not be applied directly to the vaporising chamber, which should always be surrounded by hot water. The water bath, if closed, should have a water gauge and a safety valve.

Evaporating pans: When methyl bromide is discharged into a fumigation chamber, a good practice is to let the liquid flow into a shallow evaporating pan of sufficient capacity to hold the entire dosage at a depth not greater than 12 mm (0.5 in). In this way, more even distribution of the fumigant is obtained, especially if a current of air from a fan or blower is directed across the pan; also, the danger of liquid spilling directly on the commodity is eliminated. If fumigation is being effected at temperatures below 15°C, the pan may be warmed in some way, preferably with heating coils, infra-red heaters or ordinary light bulbs.

Measurement of Dosage

Fractional measuring devices are used when the dosage of methyl bromide cannot be computed in terms of whole cylinders or cans. Sometimes the approximate dosages discharged from cylinders can be determined by weighing the cylinders on portable scales. Graduated glass measuring tubes are accurate to the nearest millimetre and are used for measuring amounts up to 280 ml (the volume of 1 lb of methyl bromide is 260 ml at 0°C).

Phosphine

Phosphine or hydrogen phosphide (PH_3) is a low molecular weight, low boiling point compound that diffuses rapidly and penetrates deeply into materials, such as large bulks of grain or tightly packed materials. The gas is produced from formulations of metallic phosphides (usually aluminium or magnesium phosphide) that contain additional materials for regulating release of the gas.

$$AlP + 3H_2O \rightarrow PH_3\wedge + Al(OH)_3$$

$$Mg_3P_2 + 6H_2O \rightarrow 2PH_3\wedge + 3Mg(OH)_2$$

Aluminium phosphide is formulated 8S tablets, pellets or small sachets of powder with additional materials such as ammonium carbamate, ammonium bicarbonate, urea and paraffin to regulate release of fumigant and suppress flammability. The magnesium phosphide is similarly manufactured in tablets or pellets. It is also prepared in flat plates; here the formulation is embedded in a plastic matrix that regulates access of moisture and hence controls release of the gas. After the phosphine has evolved from a formulation, the residue that remains consists mainly of aluminium or magnesium hydroxide. Small amounts of undecomposed aluminium phosphide may also remain in the greywhite dust from tablets, pellets or sachets.

Properties

A strong odour, resembling carbide or garlic, is normally associated with the evolution of phosphine from various formulations. It can be detected by smell even at very low concentrations. This odour seems to be due to the presence of other compounds produced along with phosphine and they may be preferentially absorbed during fumigation treatments. Under some conditions the odour may disappear, even when insecticidally effective concentrations are still present in the free space of a fumigation system. While any odour associated with the evolution of phosphine may indicate the presence of phosphine, it should not be relied on for warning purposes.

Phosphine is very toxic to all forms of animal life, hence exposure of human beings even to small amounts should be avoided. Poisoning can result from ingestion or inhalation; however, the gas is not absorbed through the skin. A concentration of 2.8 mg/l (ca 2 000 ppm in air) is lethal to humans in a very short time. The threshold limit value is usually set at 0.3 ppm for a 40-hour work week. Symptoms of poisoning for humans are described below under "First Aid". Phosphine ranks as one of the most toxic fumigants of stored product insects. It is a slow acting poison that is effective at very low concentrations if the exposure time is long enough.

Usually, exposure times of four or more days are required to control insects, depending on temperature. The toxicity of phosphine to insects declines as the temperature falls to 5°C, so that longer exposure times are required for it to exert its effect. It is not recommended for use below 5°C. The exposure time cannot be shortened by increasing the dosage; in fact, high concentrations can have a narcotic effect on insects thereby reducing mortality. Phosphine has an inhibitory effect on insect respiration and is unique in that it is only toxic to insects in the presence of oxygen—in the absence of oxygen it is not absorbed and is not toxic to insects.

However, the action of phosphine is potentiated by carbon dioxide and the exposure time can be reduced when both gases are present. Some stages of insects are considerably more tolerant to phosphine than others. The eggs and pupae are usually hardest to kill while larvae and adults succumb more easily. Reynolds et al., found that this tolerance was at least partially overcome by the development that occurred in the insects during the relatively long exposure periods.

For example, a 10-day exposure of various stages of Sitophilus Granaries was found to be long enough to permit the different stages to reach a susceptible point of development at some time during the fumigation. These results suggest that all pre-adult stages, some of which are quite tolerant to the fumigant, may reach a susceptible stage of development during a 10-day fumigation, so that an exposure period of this length will lead to complete mortality. Howe discussed problems relating to the laboratory investigation of phosphine toxicity to stored product insects.

Observations to date on the effectiveness of phosphine on mites in bulks of grain indicate that the fumigant may be successful in bringing about immediate reduction in mite populations and thus improving the condition of the grain. However, populations of some species have been seen to build up again in the grain after an interval of time. This is partially due to the fact that natural predators such as the mite, Chevletus eruditus, are eliminated and partially because some stages of destructive species of mites are resistant to the fumigant. Tests on dried prunes have shown that satisfactory control of mites on this commodity can be obtained with phosphine.

The effectiveness of phosphine can be reduced considerably by development of resistance in insects. Winks showed that Tribolium castaneum could develop 10-fold resistance to phosphine in six generations. Resistance may occur in immature stages as well as in adult insects. Bell et al., found a good correlation between resistance in the adult stage of Rhyzopertha dominica and resistance in the egg stage. Champ and Dyte found evidence of resistance to phosphine in insects from several parts of the world, particularly where inadequate techniques of fumigation were employed, and they indicated that emergence of resistance to fumigants under practical conditions was a matter of great concern.

There is recent evidence of the development of resistance to phosphine in field populations of Khapra beetle and other insects infesting stored grain. Hole discussed the variation in tolerance of seven species of stored-product Coleoptera to phosphine in strains from twenty-nine countries.

Effect on Plant

There is considerable evidence from studies so far conducted that phosphine in insecticidal treatments does not, under normal conditions, affect the germination of seeds. Strong and Lindgren tested cereal, sorghum and small legume seeds with one or two (repeat) fumigations at comparatively high concentrations. Berstlief and Alexandrescu confirmed these findings for wheat sod maize under Rumanian conditions and Pinto de Matos for the germination of groundouts.

When a third fumigation was done with phosphine, Fam et al., found no ill effect on the germination of some varieties of tomatoes, sweet melons, cucumbers, peas and beans, but one variety of broad beans was slightly affected. However, the growth and yield of plants grown from seeds subjected to repeated fumigations with phosphine may be significantly reduced. Joubert and Du Toit reported a reduction in total yield of maize grown from seed fumigated twice sod three times with phosphine.

There is little information on the tolerance of growing plants to the vapours of phosphine. Monro and Upitis found that 15 arieties of glasshouse plants were tolerant to concentrations which were completely toxic to the postembryonic life stages of the mealybug Plannococcus citri, but the eggs were not killed. The nematode Meloidopyne hapla may be controlled in potted rose plants with phosphine applied to the soil, without apparent injury to the plants.

Phosphine has been used for many years to control insects in a wide range of plant products throughout the world. To date there has been no report of appreciable adverse effects from recommended treatments. Mayr and Hild concluded that normal fumigation with phosphine has no effect on vitamins, particularly vitamins A and B_2 (riboflavin), in a group of important foods which are a regular dietary source of these vitamins. Fumigation of wheat with phosphine under normal conditions has no adverse effect on the baking quality of flour made from it. Phosphine can also be used to control insects and mites in dried fruit.

Tests on fresh fruit and vegetables show that insects such as fruit flies can be controlled using gas generated from a magnesium phosphide preparation without injury to the produce. Seo et al., found no injury on papaya, tomato, bell pepper, eggplant or banana fumigated with dosages that eliminated eggs and larvae of the fruit flies. Ten varieties of avocado, although not injured by the treatment, did ripen more quickly than unfumigated avocados. Grapefruit and tomatoes have also been fumigated without injury at concentrations sufficient to kill fruit flies.

Determination of Phosphine

For the determination of concentrations of phosphine used in fumigation, glass detector tubes are available from a number of manufacturers in various ranges from 15 to 3 000 ppm. These tubes are reasonably reliable, easy to use and are of sufficient accuracy for monitoring approximate concentrations achieved at various stages in a fumigation. For protection of personnel from low levels of phosphine around the threshold limit value (TLV), similar tubes with ranges down to 0.025 ppm can be obtained. A mixed indicator paper strip that will give rapid, sensitive and reliable detection of phosphine around the TLV has been developed by Kashi and Muthu.

Infra-red analysers have been used on an experimental basis for analysis of phosphine and found to have good stability, reproducibility and sensitivity. The concentration of phosphine is determined by measurement of absorption bands at either 4.2 μm or at 9.0 μm unless carbon dioxide is present; carbon dioxide interferes with analysis of phosphine at 4.2 μm and may need to be absorbed from the sample or the analysis may be done at 9.0 μm. In normal use, sampling can be done by pumping fumigant—air mixture through the analyses but there is also a closed loop injection system for calibration of the instrument or for experimental work.

Here small samples can be injected through a septum by a gas syringe. With the infrared analyses, concentrations up to 2.7 mg/l have been measured under field conditions and, with proper manipulation, levels down to 0.0004 mg/l (0.25 ppm) can be detected. Gas chromatography has been developed and used extensively for analysis of phosphine in experimental work. Small, portable gas chromatography suitable for analysing phosphine in commercial treatments are now available on the market. One such instrument that is simple, easy to use and virtually unaffected by air and high moisture levels can measure phosphine below 0.02 ppm.

The problem of determination of residual phosphine in fumigated foodstuffs has been reviewed by Dietrich et al. A sensitive method, developed by Bruce et al and modified by Sullivan and Murphy has been widely used for analysis of phosphine residues in many commodities. Kroeller described a sensitive method with a simplified procedure for plotting the calibration curve using high purity potassium dibydrogen phosphate. These methods are sensitive to less than 0.005 mg/kg. For analysis of desorbing residual phosphine by gas chromatography the method of Dumas can be employed.

Formulations

Both aluminium and magnesium phosphide are manufactured in several different formulations for a variety of applications. Aluminium or magnesium phosphide powder is compressed into hard round or flat tablets about 3 9 in weight or pellets of 0.6 9, which yield approximately 1 and 0.29 of phosphine, respectively. Aluminium phosphide powder is also prepared in permeable paper bags or sachets. Additional materials such as paraffin and ammonium carbamate or ammonium bicarbonate are included in the formulations to regulate moisture uptake and to dilute the phosphine as it is generated.

The products are supplied in sealed metal tubes, cans or flasks, which are packed in cases. As long as the containers remain sealed the storage life of the product is virtually unlimited. The pellets and round tablets are supplied in flasks that can be resealed after opening. For convenience and safety, pellets or sachets are sometimes supplied in predetermined quantities for specific applications. Pellets are prepared in special prepacks of 165 pellets each for treatments such as railway box car fumigations.

Similarly, sachets are joined together in ropes or in "blankets". In these prepared packages the pellets or sachets are separated sufficiently to avoid a build up of excessive heat and concentrations of the gas in small spaces. Magnesium phosphide is also marketed in the form of a flat plate about 280 x 170 x 5 mm and weighing 2069. The active ingredients of the formulation are embedded an inert polyvinyl acetate matrix fabricated in the form of a semirigid plate covered on both sides with moisture permeable paper.

Every plate is individually sealed in a gas-impermeable foil pouch, or 16 plates interconnected to form a 4 480 mm strip, are similarly sealed in foil and packaged in tins - 32 plates per tin or two strips of 16 plates each. Once the plates or strips are removed from the foil pouches, they start evolving phosphine within one half to one hour. This formulation is intended to be used for fumigation of bulk goods and packaged and processed commodities. It can be applied successfully under almost nil space storage conditions provided that the structure is tightly sealed.

According to the manufacturer, this formulation can also be used to fumigate fruits and vegetables at recommended dosages without any adverse effects, such as phytotoxicity. The plates and strips provide ease of application and collection after the treatment; there is no danger of contaminating goods with spent fumigant as the plastic matrix retains all such material. Magnesium phosphide formulations release the phosphine more rapidly than aluminium phosphide products, with the maximum gas reading usually being achieved within the first 24 hours. Magnesium phopshide made in discs weighing 109 for control of burrowing rodents and moles is for outdoor use only.

Containers of aluminium or magnesium phosphide formulations have labels that give important information concerning use, hazards and precautions. The instructions on the labels should be carefully adhered to during fumigation. Containers should be opened in the open air where any released gas can readily diffuse away.

Functions of Phosphine

Because phosphine is highly toxic, inhalation of even small quantities of the dust from the formulation, as well as the evolved yes, should be avoided. Pellets or tablets may be applied directly to a grain stream by hand or by means of automatic applicators. The rate that the yes evolves from the formulation varies, depending on type of formulation, moisture and temperature. In grain, for example, if the moisture content and temperature are high, all of the gas from aluminium phosphide formulations is evolved within three days.

Special probes are used for applying tablets below the surface of bulk grain. Sachets may be applied directly to the grain stream, pushed into the grain bulk or inserted into specially designed, permanently installed pipes in grain bins. For some treatments the sachets are laid out in blankets on the surface of grain to allow the gas to evolve and diffuse into the grain mass. For the treatment of bagged grains and other raw commodities in transport facilities, such as railway wagons, pellets or tablets may be spread evenly over the load or placed in moisture permeable envelopes to fit in some convenient location near the door before closing. When fumigating packaged commodities under gas-proof sheets the tablets or pellets can be spread out on trays to lay under the sheet before it is secured.

In warehouses, after the structure is adequately sealed, the tablets or pellets are spread out on trays or sheets of Kraft paper so that residual material can be easily collected at the end of the treatment. The tablets or pellets should never be piled on top of each other or in a mass. On completion of the fumigation, all windows and doors should be opened and the space aerated for at least two hours. A gas reading should be taken with a suitable analyses before entering the fumigated area. If it is necessary to enter the fumigated space to open doors and windows a gas mask with a canister designed for phosphine must be worn.

Hydrogen Cyanide

Hydrogen cyanide was one of the first fumigants to be used extensively under modern conditions. Its use for treating trees under tents against scale insects was developed in California in 1886. The use of HCN has been declining in recent years, but it is still important in certain fields of application. HCN is one of the most toxic of insect fumigants.

The fact that it is very soluble in water has considerable bearing on its use in practice. Thus, it may produce injury on moist materials, such as fruit and vegetables, because the solution of HCN in water is a dilute acid. Not only does this acid render these materials unpalatable and possibly hazardous for human consumption, but its action, by causing burning, wilting or discoloration, may make them unmarketable. On the other hand, HCN has been widely used for fumigating dormant nursery stock that is sufficiently dry.

It may be used for some living plants if they can be washed with water immediately after treatment to prevent burning by the acid. HCN may be employed for fumigating many dry foodstuffs, grains and seeds. Although HCN is strongly sorbed by many materials, this action is usually reversible when they are dry, and, given time, all the fumigant vapours are desorbed. With many foodstuffs little, if any, chemical reaction occurs, and there is no detectable permanent residue. Because of the high degree of sorption at atmospheric pressure, HCN does not penetrate well into some materials. It was largely because of this that vacuum fumigation was adopted.

HCN is a powerful, quick acting poison. In humans and other warm-blooded animals it induces asphyxiation by inhibiting the respiratory enzymes and renders tissues unable to absorb oxygen from the blood in the normal manner. The toxic action is reversible. In practice, this means that a person who is completely unconscious from the effects of cyanide, but whose heart is still beating, may still recover if suitable antidotes nod remedial measures are applied in time. HCN may be absorbed in toxic amounts through the unbroken skin.

Properties

Among the commonly used famigants, HCN is one of the most toxic to insects. It also has a rapid paralysing effect on most species. This action is an important consideration in dealing with insects, because sublethal concentrations may bring about apparent death. After exposure to the fumigant, the reversible action of the poison may permit the insect to recover. This reaction has already been referred to as protective stupefaction. It is important from the practical point of view because it means that the maximum recommended concentration should be attained as quickly as possible during the application of the fumigant.

The flammability limits of HCN in air lie between 6 and 41 percent by volume. These limits are well above the normally recommended fumigation doses of up to 1.5 percent (16 g/m^3 or 16 $oz/1\ 000\ ft^3$). However, it must be pointed out that, at the point of release of the gas at the beginning of a famigation, a concentration within these limits may exist for a short length of time. If there is any flame or sparks near temporary high concentrations, a serious fire or explosion could occur. In working with this fumigant, great care must be exercised to extinguish all flames and turn of all electric switches before a treatment begins.

Effect on Plant

HCN applied in dosages effective against insects does not affect the viability of seeds that are normally dry, with moisture contents suitable for storage. In a comprehensive study, the results of which were published between 1959 and 1961, Strong and Lindgren studied the effect of HCN in insecticidal concentrations on the germination of a wide range of seeds, including grain seeds, flax and small legumes. The variable factors considered were moisture content, repeat fumigations and postfumigation storage. It was concluded that germination of wheat, barley, oats, rice and flax seeds was not impaired by one or two famigations with HCN.

With the small legumes with a range of 5.8 to 12.2 percent moisture content, all were tolerant to HCN in one or two fumigations, except a like clover for which there was positive indication of impairment of germination. Among 80 varieties of grain, vegetable and flower seeds tested by Lindgren et al, six showed evidence of reduced germination. These were pole beans, burnet, California black mustard, smilo, marigold and snapdragon. It may be concluded that HCN is a safe fumigant to use for seed treatment, especially for cereal grains under normal conditions, but with flower and vegetable seeds preliminary trials with local varieties are advisable.

A considerable amount of injury, either temporary or permanent, may be sustained by actively growing plants fumigated with HCN. Because this gas is very soluble in water, special precautions have to be taken to reduce the amount of moisture on leaves and stems and in pots or soil balls of actively growing plants. Therefore, the plants should not be watered for one or more days before treatment. After treatment with HCN, it is necessary to wash the plants with water to remove any residual acid.

HCN interferes with photosynthesis and other physiological processes; for that reason, plants may be more susceptible to injury in daylight. Usually, treatments have to be carried out at night or in the dark. Also, following exposure, plants should be kept away from sunlight for several hours. In the past HCN was widely used for fumigating ornamental and glasshouse plants, but it has been replaced by other fumigants that are less phytotoxic. HCN was extensively used for may years to control scale insects on citrus trees in a tent fumigation procedure. In this treatment, HCN gas (5 g/

m) from liquid HCN, or evolved from a salt such as calcium cyanide, was liberated into a relatively gas-tight tent installed over the tree and maintained for usually around 45 mintues.

A number of factors, including temperature, humidity, physiological conditions of the tree, have considerable influence on the tolerance of the trees to the fumigant. Since different species or varieties of citrus trees show wide variation in response and the insects in different areas may vary in tolerance, the actual conditions of treatment usually have to be developed to suit local circumstances. In scale insect eradication work, where complete kill of all insects is essential, the tent fumigation treatment with HCN has been found to be a valuable technique.

HCN was used by Del Rivero et al, to control larvae and eggs of the woolly white fly on orange trees. While the concentrations of HCN required to kill insects may cause severe injury to growing plants, lower concentrations of HCN may stimulate growth and be useful in "forcing" growth in some species.

HCN is still used in many parts of the world for the fumigation of the dormant nursery stock of deciduous trees, especially in quarantine treatments against the spread of scale insects. As in growing plants, a marked stimulation has been noticed in some nursery stock fumigated with HCN. Immediately following fumigation it is sometimes necessary to wash plants with water to avoid the burning of buds and new foliage.

For the most part, HCN does not react with the constituents of foodstuffs of normal moisture content. For example, materials such as honey and beeswax that have been exposed to HCN in treatments designed to destroy the bee colonies at the end of a season, were found to have less than 0.004 and 0.02 mg/kg CN.

Effects on Fruit and Vegetables

HCN is not generally recommended for the fumigation of fresh fruit and vegetables. It is used to some extent in the treatment of citrus and other fruit for the control of scale insect or thrips. In some countries HCN is used in quarantine treatments of bananas, pineapple and other commodities for the control of aphids, mealy bugs and other exposed insects. The European Plant Protection Organisation recommends HCN in a vacuum treatment for bulbs, rhizomes, tubers, asparagus roots and strawberry plants to control certain mites and nematodes.

Effects on Cereals

HCN has been used safely and successfully against a wide range of dry plant products used as foods. An interesting exception to the general rule is that some HCN is retained by fumigated bran in combined form, hut this residue has no adverse effect on mixed feed containing average amounts of the treated bran. At atmospheric pressure, it does not penetrate as quickly or as effectively as dues methyl bromide. This handicap can be overcome in bulk grain fumigations by the use of recirculating systems. Kunz et al., studied the penetration of HCN through a mass of grain sorghum. The variables of temperature, moisture and dockage all reduced penetration in proportion to their increase.

Effects on Dried Fruit

Although HCN has been used for many years in the treatment of dried fruit, this practice is safe only under carefully controlled conditions. HCN may react with laevolose to form laevolose cyenohydrin, a poisonous compound that is fairly stable. The amount of cyanohydrin formed in dried fruit varies with its moisture content and this should be kept to a minimum during fumigation.

With properly conducted treatments at normally recommended dosages and exposure periods, the small amounts of residues that may form would not present a danger. In one instance, careless application of the fumigant permitted liquid HCN to run into packages of dried fruit and an outbreak of food poisoning ensued from the excessive residues of hydrocyenic acid present at the time of consumption.

Determination of HCN

Various devices are available for determining whether a space fumigated with HCN is sufficiently aerated to permit human entry or whether a given material is adequately free of gas to allow handling. An important use of these devices is for testing bedding and clothing after fumigation in order to discover whether HCN is sufficiently desorbed so that warmth from the body may not release dangerous concentrations of the gas. All "handy" devices for indicating fumigant concentrations should be used with care. Instructions for their use should be followed meticulously. Judgement is required for the proper interpretation of the readings, and this improves with experience.

Test papers: Specially prepared test papers have been used successfully as indicators of the presence of HCN in dangerous concentrations. At present, two kinds are in general use, referred to respectively as methyl orange and benzidine copper acetate test papers. Filter papers dipped in a solution containing methyl orange and mercuric chloride give reliable indications only down to 25 ppm of HCN in air at temperatures above 10°C. However, if kept moist in a tightly stoppered bottle away from light, the papers will be in a fit condition for practical use for up to three months after preparation. The reaction of the papers is slow.

In the presence of HCN they turn from an orange colour to various shades of pink, the usual period of exposure being two minutes. It is unnecessary for the observer to Stay in the space being tested; often the papers may be lowered into a space on a string or held through a door or window on the end of a stick. The papers made from a solution containing benzidine acetate and copper acetate are sensitive to 10 ppm of HCN in air. The change in colour from white to various shades of blue, according to the concentration of HCN, takes place in ten seconds, and it is usually necessary for the operator to make the reading in the contaminated atmosphere. These papers have to be dipped in the mixture of solutions immediately before the test is made. Also, because the mixture is very unstable, the two solutions must be kept separate and only mixed shortly before use.

The benzidine acetate solution itself is unstable and must be kept away from light in a brown bottle. Since benzidine compounds are known to have carcinogenic potential, appropriate precautions should be taken if this test method is used. Manufacturers or suppliers of various types of cyanide fumigants give full instructions for making and using test papers. Some manufacturers also supply the methyl orange test papers ready for use and others have special test kits available for making the benzidine copper acetate papers on the job.

Automatic toxic gas detector: A portable, fully automatic detector based on the colour stain formed by reaction of the gas with reagents in a chemically impregnated paper is available for measuring hydrogen cyanide at TLV levels. This instrument provides in a rugged, lightweight unit, a sampling pump, sampling head and a simple photooptic system for direct reading of the gas concentrations. By changing the sampling period and/or the electronic gain, a whole range of concentrations can be analysed. Tapes are available from the manufacturer for a number of different toxic gases.

Glass detector tubes: A number of glass detector tubes are available on the market for detection and measurement of HCN. Low range tubes that give concentration readings from 1 - 50 ppm can be used for health protection purposes and high range tubes designed to measure the levels used in fumigation treatments (0.05 to 2.0 percent) are available.

Chemical analysis: A simple method for determining HCN concentrations during fumigation is the silver nitrate method of Liebig, with modifications. Samples of air/gas mixture are removed by aspiration and the HCN is absorbed in 2 percent sodium hydroxide solution. Titration is carried out with a standardised silver nitrate solution in the presence of potassium iodide.

White described a simple field method based upon the established phenolphthalein test for HCN. The air/gas sample drawn from the fumigation system is scrubbed through a trisodium phosphate solution containing phenolphthalein and copper sulphate and the fumigant concentration is measured calorimetrically. The use of the thermal conductivity analyser with this fumigant is not recommended by Heseltine because of a number of technical difficulties involved.

Laboratory analysis: HCN can be measured over a range of concentrations by gas chromatography.

Application

HCN is generated and applied in several different ways;

the choice of method depends on convenience, cost and the type of structure or material being treated.

Generation by Acid on Salt

HCN for fumigation was first generated by the action of an acid on a cyanide. Although largely replaced by other more convenient techniques, this reaction is still popular for certain purposes because it is economical and lends itself to emergency treatments. Both potassium cyanide, KCN, and sodium cyanide, NaCN, may be used. Sodium cyanide is cheaper and is the salt used principally today. Sodium cyanide for fumigation should contain not less than 96 percent NaCN and is best purchased in moulds each weighing an amount convenient for measuring dosages. Sulphuric acid, H_2SO_4, is the only commercial acid which gives a good reaction without undesirable side effects. It should be purchased in the commercially pure form at 66° Beaume'.

Generation from Calcium Cyanide

Calcium cyanide, Ca $(CN)^2$, is a dark grey powder which yields HCN in reaction with the moisture in the air. The material is marketed, sometimes with a carrier, in different degrees of fineness of the granulations according to the intended use. For example, a fine grade is used for blowing small amounts into ants' nests with a hand duster. Larger granules are used for application to soil, for spreading on the floors of glasshouses or for mixing with grain. The amount of gaseous HCN yielded is between 25 and 50 percent of a given weight of granular material.

Vocalisation of Liquid HCN

HCN may be purchased in cylinders as a liquid of 96 to 98 percent purity. The liquid contains a stabiliser which extends the storage life of the fumigant by inhibiting its tendency to generate heat during decomposition and to explode. Nevertheless, the manufacturers place a limit on the length of time a given lot may be kept, usually six months, after which the cylinder must be returned to the factory. This time limit should be strictly adhered to.

At normal temperatures, the natural vapour pressure of the fumigant is not sufficient to propel it into the open, especially when the liquid has to flow through piping. It is necessary to apply compressed air above the surface of the liquid to force it through the siphon tube out of the cylinder. In vacuum fumigation, the reduced pressure in the chamber is sufficient to draw the liquid HCN from the cylinder.

The discharge tube should pass through a heat exchanger en route to the vacuum chamber, as rapid volatilisation of the fumigant brings about loss of heat with the possibility of freezing in the lines and tubes. The USE of the liquid in cylinders is a more expensive method of dispening the fumigant but, in large-scale operations, it is much safer and more convenient to handle HCN in this form and lengthy postfumigation cleaning is eliminated.

Absorption

HCN may be purchased absorbed in inert materials. A convenient method of absorption is the use of highly porous cardboard discs. They are shipped in carefully sealed tin cans that withstand the pressure exerted by the absorbed fumigant at the highest temperatures normally encountered. Before the cans are opened, it is advisable to precool them in cold water, ice, or in a refrigerator at 0°C. They are opened with special can openers supplied by the manufacturers, and the discs are scattered in the space to be fumigated according to the required dosage.

It is necessary that proper respirators be worn by those opening the cans and distributing the discs. The discs are sometimes marketed with 2 to 5 percent chloropicrin used as a warning agent. Small amounts of ethyl bromoacetate have also been used for this purpose.

Minor Fumigants

Acetaldehyde

Acetaldehyde is a naturally occurring volatile compound that is toxic to insects as well as certain fungi, bacteria and yeasts. Tests on fruit and vegetables have shown that it can be used to control some insects, such as aphids and thrips, without undue injury to the fresh produce. Concentrations up to 3 percent for 30 minutes have been applied to apples as a fungicidal treatment without causing injury to the fruit. The boiling point of acetaldebyde is 21°C, it has a characteristic pungent odour and is flammable at concentrations of 4 percent or greater.

Acetaldehyde is less toxic to mammals than other commercial fumigants, such as ethylene dibromide. It has a narcotic action on the nervous system and causes irritation to the eyes and mucous membranes. Large doses may cause death by respiratory paralysis. This compound is suspected of having carinogenic activity in humans; however, it is not likely to be hazardous as a residue on fresh produce because it does not accumulate in tissues.

Plant tissues may absorb acetaldehyde but it is metabolised to acetic acid, ethanol and carbon dioxide. Acetaldehyde occurs naturally in fruit and vegetables, it is used as a flavouring agent and has been registered as a food additive in the United States. This compound and other naturally occurring volatile compounds, such as ethyl formate, may have some potential as future replacements for fumigants that leave harmful residues.

Azobenzene

This compound (boiling point 293°C) is a crystalline solid which has proved useful in the past for the control of mites in greenhouses. It acts as a vapour formed by heating the pure crystals by means of steam pipes, hot plates or lamps; it is also evolved by igniting a powder in a pressure fumigator. Many glasshouse plants and blooms are tolerant, but there has been some discoloration of red flowers. Flowers of the African violet (Saintpaulia) are reported to be susceptible to damage.

Chloroform

Chloroform (boiling point 61°C) is not highly toxic to insects but it has shown considerable promise as a constituent of liquid-type fumigants, in which it serves as a carrier for other more toxic ingredients, such as ethylene dibromide and carbon disulphide. Chloroform is nonflammable when mixed in air in any proportion. However, it is listed as being a compound which is "suspect of inducing cancer" by the American Conference of Governmental Industrial Hygienists.

Dichloronitroethane

This fumigant (boiling point 124°C) was introduced some years ago and appeared useful for treating grain, other stored products and soil. It gives warning by odour and irritation of the eyes. It may be strongly corrosive to metals in moist atmospheres.

Ethylene Chlorobromide

Ethylene chlorobromide is effective against the oriental fruit fly, but it is not as toxic as the closely allied ethylene dibromide. It has a lower boiling point than EDB. The effect of this fumigant on fruit and plant material is similar to that of EDB. It has not come into general use, although it has been adequate in many fields of application for which EDB is also suited.

Methallyl chloride

This gas, also called methallyl chloride, has been suggested as a grain fumigant and has been used experimentally in mixtures with other fumigants. In laboratory tests it has ranked above carbon disulphide and ethylene dichloride against eight species of stored-product insects and it is considerably more toxic than carbon tetrachloride, although less toxic than ethylene dibromide.

Methallyl chloride has a boiling point of 72°C and is nonflammable in concentrations toxic to insects. In field trials in Russia it has shown promise as a fumigant for cereal grains and pulses and is significantly less expensive to apply than chloropicrin and methyl bromide. Methallyl chloride also shows promise for fumigation of individual sacks of grain.

Methylene Chloride

Methylene chloride has been found useful as an ingredient of fumigant mixtures. It is nonflammable and has a boiling point of 40.2°C. In itself it is toxic to insects, but ranks low in toxicity in comparison with other commonly used materials.

Nicotine

Nicotine (boiling point 247°C), vaporised in different ways, was formerly used extensively as a glasshouse fumigant. The most convenient method of volatilising it is by the use of pressure cans; these are ignited to send off dense clouds of smoke containing the vapours of nicotine. In a glasshouse, it was found that the nicotine concentrations fell off rapidly and little or none remained one hour after discharge.

When the nicotine was evaporated slowly, the concentration was maintained only as long as evaporation continued and the introduction of fumigant was equal to the loss through leakage. Delicate flowers and tender shoots may be injured by nicotine applied in any way.

Propylene Dichloride

In empty containers, this compound compares favourably with many other fumigants in toxicity to insects. It has been tried as a grain fumigant, usually in mixture with carbon tetrachloride, but so far has not been widely adopted.

Sulphur Dioxide

Sulphur dioxide (SO_2) should be mentioned if only to point out its disadvantages. It is the oldest fumigant known to man, having been used from time immemorial by burning sulphur. Nowadays it is sometimes used as an ingredient of liquid grain fumigants. Intrinsically it is quite toxic to insects. It also serves as a warning gas in these mixtures on account of its intensely irritating properties to humans. It is rapidly sorbed by any material undergoing treatment; it has a deleterious effect on grain and flour and is highly corrosive to metals.

Carbon Dioxide

Carbon dioxide (CO_2) is an ingredient of our normal atmosphere and is not to be considered as a poisonous fumigant in the ordinary sense. Nevertheless, it has an important role in insect control.

3

Fumigant Mixtures

Fumigants are sometimes marketed in mixtures with other compounds. There are a number of practical reasons for combining fumigants, the more important being:

1. The flammability risk of the toxic ingredient is reduced or prevented altogether by the addition of another chemical. Examples are the addition of carbon dioxide to ethylene oxide or methyl formate and of carbon tetrachloride to acrylonitrile or carbon disulphide.
2. With liquid-type fumigants in grain fumigation, mixtures may be made to provide ingredients whose vapours have differing rates and patterns of diffusion. After the liquid has been applied to the surface the ingredients evaporate. The distance the ingredients diffuse downward depends largely on the extent to which they are sorbed by the grain. It may, therefore, be necessary to have various types of ingredients in the mixture to kill insects at various depths and locations in the grain. An example is a mixture of ethylene dibromide, ethylene dichloride, carbon disulphide and carbon

tetrachloride, which is used for grain fumigation in farm storage units and country elevators.

Similarly with gaseous fumigants, mixtures may be used to obtain complementary advantages. Phosphine is used in combination with methyl bromide in treatments of materials such as oil-seed expeller; the phosphine penetrates deeply into the material while the methyl bromide gives more control on the outer surfaces.

3. A highly volatile fumigant such as methyl bromide may diffuse downwards too rapidly, so that the upper part of a load of infested goods may not receive an adequate insecticidal treatment. Another less volatile fumigant, such as ethylene dibromide, is added to ensure that the material at the top is properly fumigated. A mixture of this type may be useful under tropical conditions when commodity temperatures may range as high as 38°C.
4. The principal toxic ingredient may be diluted so that its distribution becomes more uniform. Carbon tetrachloride, though only moderately insecticidal in itself, aids in the distribution of other fumigants, such as ethylene dibromide, which does not diffuse well through masses of grain.

Although fumigant mixtures are very useful for many types of application, especially for the treatment of grain in bulk and for local or spot fumigations, it should be pointed out that their use involves certain complications. As indicated above, the ingredients may settle out in different parts of the fumigation system. This may result in a situation whereby the more toxic material is acting against one part of the insect population while the less toxic ingredients are acting against the remainder.

Erratic results may thus be recorded unless full provision is made to allow time for the less toxic materials to exert their maximum effect. Another objection, which is possibly more serious, is the fact that commonly used methods of field analysis of vapours, such as the thermal conductivity analyses, will not differentiate between the components of the mixture so that is not practicable to attempt to keep a check on the effective distribution of the toxic ingredients. Any reading taken with this type of instrument is meaningless under these conditions.

A gas chromatograph or an infrared analyser would provide the necessary information in the field, or samples could be brought to the laboratory for immediate analysis. Kenaga has suggested the use of the mass spectrometer for the analysis of mixtures but this expensive instrument would have to be used in the laboratory.

Acrylonitrile

Because of its flammability, acrylonitrile is never applied alone but always in admixture with nonflammable materials. Mixed with carbon tetrachloride it has been used as a local fumigant and for the fumigation of tobacco. Mixed with chloroform or methylene chloride it has been employed with some success for fumigating buildings against dry wood termites. Although in comparison with methyl bromide these mixtures do not have a deleterious effect on household materials, such as sponge rubber, they do not aerate after fumigation as rapidly as methyl bromide or sulphuryl fluoride. Ruppel et al., concluded that mixtures of acrylonitrile with carbon tetrachloride were not fully satisfactory against pests of stored maize in Colombia as compared with the EDC : CT 3 to 1 mixture commonly used in that country.

Carbon Disulphide

In mixture with carbon tetrachloride, carbon disulphide (CS_2) is very useful for grain fumigation, especially in tall, upright silo storage units and in all storage units equipped with adequate recirculation systems. Enough carbon tetracilloride should he incorporated in the mixture to eliminate the fire hazard. A mixture commonly marketed for these purposes contains, by volume, 16.5 percent CS_2 82.5 percent carbon tetrachloride and 1.0 percent inert ingredients. Another mixture useful for gravity distribution in concrete or metal grain bins contains 24 percent CS_2, 71 percent chloroform and 5 percent ethylene dibromide by volume.

Chloropicrin

The highly insecticidal and lachrymatory properties of chloropicrin have been put to use by mixing this chemical with other materials for the fumigation of grain in farm or bulk storage units. It has been marketed in a mixture with methyl bromide or carbon tetrachloride. A mixture of chloropicrin and methyl chloride 85 : 15 is recommended for forced distribution fumigation of grains.

Ethylene Dibromide

Ethylene dibromide (EDB) is an ingredient of many important fumigant mixtures. It is highly insecticidal, and under normal conditions the residues remaining in foodstuffs are of a low order. These advantages are offset by the fact that EDB is considered to be very hazardous to mammals and is highly sorbed by materials undergoing fumigation. As a result, there may be poor penetration during actual treatment and prolonged persistence of the vapours in the fumigated commodity during the aeration process. Attempts to utilise the advantages and overcome the disadvantages have resulted in the formulation of a wide variety of mixtures.

Liquid formulations: Alone or with other toxic fumigants EDB is commonly mixed with carbon tetrachloride (CT) as a grain or local spot fumigant. The inclusion of CT as a high proportion of the formulation appears to add greatly to the effectiveness of EDB, principally because CT acts as an eluant for EDB in a column of grain and assists effectively in its downward migration. The following are some liquid formulations containing EDB in admixture with other materials (percent by volume) that have been widely used:

Grain Fumigants

EDB 5:	CT 95 - General grain fumigant
EDB 7:	Ethylene dichloride (EDC) 29: CT 64 - Gravity distribution grain fumigant
EDB 7:	Carbon disulphide (CS_2) 12: CT 81 Gravity distribution in flat storages
EDB 3.5:	EDC 10: CS_2 10: CT 76.5 - Gravity distribution in flat storages
EDB 5:	CS 24: Chloroform 71 - Gravity distribution grain fumigant

Local (spot) fumigants

EDB 20:	EDC 20: CI 60
EDB 59:	EDC 9: CI 32
FDB 15:	CT 85

Ethylene Dibromide-methyl Bromide Mixtures

Mixtures of EDB with methyl bromide have been found useful for three main purposes:

- in the tropics to treat bagged plant products under gas-proof sheets;

- as local fumigants for treating food handling equipment in mills and food processing plants generally;
- as a fumigant for grain stored in bulk.

Bagged goods: In India, mixtures of the two fumigants in various proportions have been recommended for application under sheets or in chambers to treat different types of commodities in bags under warehouse conditions. Treatment with the fumigant mixtures accompanied by simultaneous prophylactic applications of liquid insecticides to prevent reinfestation is known as the "Durofume" process. These authors give detailed recommendations for the proportions of the two chemicals required to achieve the best results with different materials.

A special applicator for administering fumigant mixtures of EDB and methyl bromide, which is particularly suitable for treating stacked commodities under gas-proof sheets, has been described by Majumder et al. This consists of a brass or stainless steel tube held horizontally, containing the EDB. Above this there is a vertical scaffold into which a can of methyl bromide may be inserted. The methyl bromide can is pierced at the bottom by a probe, which connects to one end of the tube containing the EDB. The methyl bromide with its natural vapour pressure forces the EDB out through a discharge nozzle at the other end of the horizontal tube so that the mixture may be led through a distribution system above the stacked commodity.

Local fumigants: A mixture of seven parts of EDB and three parts of methyl bromide has found wide use as a local or spot fumigant. This is best introduced by special applicators into holes bored at strategic points in the food handling equipment.

Bulk grain fumigation: The mixture has been found effective for bulk grain. The proportions of the two fumigants may vary according to the material being treated and the method of application. For cereal grain treatments the ratio of 70 : 30 EDB to methyl bromide is recommended for gravity distribution in smaller bins or farm storages. For forced distribution systems, 30 : 70 EDB to methyl bromide is used. Majumder et al, found that the presence of methyl bromide improved the penetration and distribution of EDB in columns of grain and milled materials, the most effective proportions varying according to the material under treatment.

Dawson has taken out a United States patent for a yelled fumigant composition made by agitating different amounts of methyl bromide and ethylene dibromide in mixture with 3 to 5 percent by weight of colloidal silica. The gel thus produced retards the evaporation of the constituent fumigants and thus affords a convenient method for applying them to grain.

Insect toxicity: From the limited amount of laboratory evidence available to date, it appears that mixtures of these two fumigants are intrinsically more toxic to stored product insects than methyl bromide alone and at least equally as toxic as ethylene dibromide alone. These authors also concluded that the mixtures killed the postembryonic stages of Tribolium confusum more rapidly than either fumigant alone.

Penetration of flour: Heuser studied the pattern of diffusion through flour at 25°C of the two components of a 1:1 by weight mixture of EDB and methyl bromide. He found that the methyl bromide components in insecticidal concentrations could penetrate to 18 in (45 cm) after 48 hours, but EDB only penetrated effectively to 2 in (5 cm) after the same time. He concluded that if EDB is used "in hightemperature zones to produce a longer lasting insecticidal effect in the surface layers than is obtained with M. B. (methyl bromide) for example where the gasproofing of fumigation sheeting is suspect, then its addition should be regarded as a supplement only, and not as in any way replacing a proportion of the methyl bromide dose".

Ethylene Dichloride

Compared with most modern fumigants, ethylene dichloride (EDC) is intrinsically not very toxic to insects but its mixture with CT, usually in the proportion by volume of 3 : 1 EDC to CT, has been used throughout the world as a successful fumigant for stored grain in a variety of structures. It has proved effective under tropical conditions in Africa. Its use for this purpose is contingent on a long exposure period, usually not less than 14 days, before the grain is turned or the fumigant aerated. Such long exposure periods do not appear to have adverse effects on the grain. It may be applied either by gravity distribution or forced distribution systems. This mixture has also been widely used as a seed fumigant. Properly used it does not have an adverse effect on the germination of seeds.

Methyl Bromide-phosphine

A method for the combined use of methyl bromide and phosphine has been tested for treatments in oil-seed expeller, a dense material in which methyl bromide does not easily penetrate. The phosphine is found to diffuse into the expeller to give control while methyl bromide is effective on the outer surfaces. Furthermore, mixtures of methyl bromide and phosphine have been shown to have a joint action that improves the insecticidal effectiveness of both compounds.

In this treatment, the phosphine-producing formulation is distributed evenly over the surface of the goods and, after sealing, the methyl bromide is applied through a tube by a double jet to vaporise in the Free space above the goods. Dosage of 56 g/m^3 methyl bromide and 4.4 g/m^3 phosphine at 29.5°C for 72 hours gave satisfactory control of Khapra beetle larvae.

Methyl Chloroform

Methyl chloroform (1,1,1 - trichloroethane) has been tested under field conditions in mixtures with ethylene dichloride and the trials have shown that the two compounds interact physically to improve distribution of each component through grain bulks. When applied with methyl chloroform, EDC is carried to a greater depth than when used alone, and the two compounds seem to act jointly to give a much enhanced toxic effect on insects. The physical properties of methyl chloroform relating to distribution and persistence compare favourably with carbon tetrachloride as does its toxicity to insects and mites.

The available toxicological evidence suggests that it is a much safer compound than carbon tetrachloride for a pest control operator to use and tests on residues suggest that it should not be a toxic hazard in grain. The potential of methyl chloroform as a substitute for carbon tetrachloride in fumigant mixtures may be indicated by further testing in the future.

4

Chamber Fumigation

Space fumigation is convenient for the designation of a wide range of treatments in enclosed spaces, which either contain infested materials or residual insect populations. The calculation of the dosage of fumigant to be applied is based primarily on the volume of the space. Fumigation may be successfully carried out in any structure that can be made sufficiently gas-tight for the length of time required. Infestations may often be dealt with on the spot without the necessity of moving the affected material.

Sealing Techniques

Since sealing techniques are common to all types of fumigation. In practice, the choice of materials is largely influencer by their availability.

Cracks and Holes

Masking tape, heavy kraft paper applied with flour paste, caulking compound, paint-on adhesives, froth packs of Styrofoam, etc. may be used to cover or seal small cracks and holes. A vinyl plastic sealing compound is available that can be sprayed over all holes, cracks and crevices. This method, sometimes called "cocooning", is effective but requires a compressor for air, a spray gun and a hose.

Instructions for carrying out the process are provided by manufacturers of the plastic.

Heavy kraft paper applied with flour paste and impregnateti with a heavy grease or sheet of polyethylene sealed with masking tape may be used for larger openings. The vinyl sealing compound described above in the cocooning method can be applied to bridge larger openings and it will provide a relatively gas-impermeable, durable seal.

Ventilators

Heavy kraft paper can sometimes be used but polyethylene sheeting is relatively inexpensive and can be re-used several times. It can be tied with ropes to make a tight seal around a ventilator.

Chamber Fumigation

To conduct fumigations of commodities on a nonemergency basis, it is advisable to install a specially designed fumigation chamber. For many commodities the infestation may be controlled by fumigation at atmospheric pressure. Vacuum fumigation is recommended for the treatment of certain densely packed or absorbent materials. It is also used when a rapid turnover of certain goods is required. The purpose of a fumigation chamber is to allow fumigations to be carried out efficiently, safely and economically. The basic elements for design and construction should be incorporated in all chambers with variations made to suit individual needs. An effective fumigation chamber must be:

- soundly constructed so as to be gas tight;
- provided with an efficient system for applying and distributing the fumigant;

– provided with an efficient system for removing fumigant at the end of treatment;
– sited so as to handle infested goods conveniently;
– sited and operated so as to present no hazard to personnel working with or near the chamber.

There is room for very wide variation in the design of atmospheric fumigation chambers. The fumigant to be used, the siting and size of the chamber, the type of goods and methods of handling, the availability of building materials, as well as the expenditure that can be allowed for increased efficiency and ease of operation, must all be taken into consideration. If chambers are used constantly or are loaded with heavy materials, a sturdy construction is necessary.

However, light, portable chambers made from plywood or similar material may be appropriate for some types of treatment. A portable gas-tight tent was devised by Brown and Heseltine and a full description of its construction and operation was given by these authors. It proved to be very useful for the fumigation of small commodity lots and for nursery stock.

While the safest arrangement is to have the fumigation chamber located outside the main buildings, there may be circumstances in which it is safe to have it inside. If the chamber is properly constructed and equipped, it may be placed in a well-ventilated part of a building not regularly used by employees. When a chamber is located where people are constantly working it may be separated from the working space by an additional wall with a ventilated space between. The cost of loading and unloading the chamber largely determines the cost of fumigation. Therefore, by choosing 8 site which will permit the handling of goods in an efficient and economical manner, these costs can be reduced to a minimum.

After the removal of treated materials from a chamber, fumigants continue to diffuse. Therefore, the commodity must be kept on an open platform or in a well-ventilated room for 24 hours after treatment. One important way to protect employees from exposure to fumigants is by means of a well designed exhaust system. This system should be equipped with one or more fans or blowers powerful enough to draw a continuous draught of air through the door into the outside atmosphere.

A movement of air equivalent to one complete air change every one to three minutes in the empty chamber is recommended. If the chamber is located near dwelling houses, the exhaust pipe should terminate well above the roof of the building so that fumes may not enter occupied rooms during the aeration process. In some countries the spent fumigant is absorbed by passing the exhaust gases through absorbing solutions and degrading to harmless end products. For chambers up to 56 m^3 in capacity (2 000 ft^3), an exhaust stack reaching 4 m (13 ft) above the roof or any nearby structure should be adequate.

When fumigants are thoroughly mixed with air, they dissipate rapidly on discharge into the open air from the fumigation chamber. If the chamber is to be permanently fixed inside a building, it may incorporate a part of the floor, two existing walls and even the ceiling. Suggestions for constructing a chamber from various materials, together with details of accessory equipment, are given below.

The chamber size should be such that it can normally be loaded to its full capacity. There is, however, no difficulty in fumigating a partly filled chamber. While the dimensions of a fumigation chamber can only be decided by the owners, as a general guide a chamber should be approximately twice as long as it is wide with a height of 2-3 m. Even distribution of fumigant is more easily obtained in such a chamber than in a square one or one more than 3 m high.

The maximum economic size is probably of the order of 100 m^3. If a greater capacity is required, two or more smaller chambers often prove more economical and easy to handle. One chamber can be unloaded and reloaded while the second chamber is in use. The dimensions of the chamber are as important as the overall capacity, particularly where mechanical handling machinery is used for instance, the size of pallets, if used, will dictate the proportions of the length and width. The height will depend on whether goods are loaded manually or with a fork lift truck.

The most satisfactory type of chamber, and one which is likely to give the minimum of trouble from leakage, is one with a concrete floor, walls of brick or poured dense concrete or other similar solid building materials and a flat roof of reinforced concrete. Chambers made of timber framing covered with sheets of material are likely to produce initial difficulties in sealing and continuing difficulties due to deterioration or damage. When made of light materials, chambers are prone to leakage and may need an impermeable flexible film lining on the inner surface or regular renewal of the painted finish.

Paints should be applied, preferably by spraying, to give a continuous, impervious surface. A relatively durable finish can be obtained with an alkali-resisting primer followed by two coats of black oil paint and two coats of white. The use of the two colours facilitates inspection of damage after the chamber has been used.

Chamber Construction

Walls: The outside walls of the chamber may be constructed wholly or in part of concrete, concrete blocks, sheet iron, plywood or tongue and groove boards. All these materials may also form part of the inner lining, but careful sealing of all joints and seams is essential. Rough concrete and brick sorb fumigants and, if used, must be covered with hard finishing cement, with two or three layers of asphalt paint applied over its surface.

Plywood sheets, held by a framework of 5 x 10 cm (2 x 4 in) timbers, may be used for the interior walls of chambers. All seams between sheets and at junctions of the floor with walls and ceiling must be carefully sealed with materials such asphalt cement or similar caulking compounds. With plywood it is advisable to seal all the inside surfaces with a primer, which is then covered with a suitable resin-base varnish. This protects the wood from moisture and reduces loss of fumigant through sorption by the wood. If a plywood-lined chamber is to be subjected to rough usage with heavy bays or cartons, it is suggested that a sheet-metal lining be added.

All joints and nail holes in the sheet metal should be carefully soldered. When tongue and groove boards are used, they must be in two layers, placed diagonally, with a layer of roofing paper between. Insulation may be pieced between the layers if required. All joints must be sealed with great care, using a suitable cement. The inside surface should then be covered with a plastic film lining or with a primer and varnished as recommended above for plywood. If a chamber detached from a building is to be used in the winter, it is recommended that the outer walls be of cement or brick.

The inner wall may be made of plywood, which is separated from the outer wall by insulating material and a vapour barrier. Careful attention must be paid to the proper sealing of the plywood to prevent an accumulation of fumigant in the interspace. Sheet or corrugated iron has been used successfully in warmer climates for the walls of chambers. When metal is used, great care must be taken to ensure proper sealing.

All seams and joints are liberally filled with mastic end the edges of the overlapping sheets covered with mastic tape. In chambers of this type, the ends, top end bottom of the corrugated iron well rest firmly in neoprene gaskets. It is necessary to have more support members than in a standard sheet iron building, in order to reduce expansion end contraction.

Ceiling: Lighter gauges of the materials recommended for the walls may be used for the ceiling. With plywood, a thickness of 1 cm (0.4 in) is usually sufficient. A detached permanent chamber must have a roof above the ceiling to provide all-year-round protection from the weather.

Floor: When the chamber is in constant use, and especially for treating bagged goods, a concrete floor is best. This should be reinforced to bear the greatest expected load. The concrete must be hard-finished to provide a gasproof, nonabsorptive surface. Tongue and groove lumber or plywood, satisfactorily sealed as described above, may be used if built to withstand the loads. Plywood flooring of 1.2 cm (0.5 in) is often suitable for plants and nursery stock if it is well finished in the way described above.

Doors: The chamber can be provided with one or two doors, depending on its size and function. For larger chambers the use of two doors allows the loading of untreated material in one end of the chamber with unloading after treatment from the opposite end. If a chamber is located inside a building, it may be built through a dividing wall with loading and unloading doors on opposite sides of the wall. The segregation of untreated and treated stock in this manner reduces the possibility of crossinfestation.

The door should be as light as possible, it must give a gas-tight fit and it should be of good quality to withstand constant use. A stiff steel or timber frame clad with sheet metal on the inner surface may be satisfactory. The size of the door will be governed by the requirements for loading and unloading, but the smaller the door the easier it is to ensure a gas-tight fit. Lift doors, which are raised above the chamber during loading and unloading, are very practical when a chamber is inside a building, because they are out of the way when open and do not interfere with the movement of the goods.

Lift doors are operated by means of a counterweight and, if this works properly, opening and closing are easy. Doors sliding on rails are better for installation on the outside of a building, where they can be rolled out of the way along an outside wall. An ordinary hinged door may be installed in a chamber if the materials to be treated can be moved in and out easily. Generally, the doors may be made of the same wood or metal materials, suitably finished, as recommended for the walls.

Door gasket: Proper closure is important because fumigants, such as methyl bromide, escape readily through improperly sealed doors. The best seal is provided by a continuous strip of rubber or neoprene that is soft enough to give a good seal but resilient enough to recover well after continued pressure. Natural rubber may deteriorate after some time if used with methyl bromide, but is readily renewed. Overhead doors or hinged doors should fit neatly into flanges.

The best seal is obtained with three strips of rubber, two on one surface, with a gap between that is just wide enough to receive the third gasket, which is on the opposite face. Sliding doors will carry only one strip of gasket on one surface, and this should therefore be 2.5 to 5 cm (1 to 2 in) wide. With all types of doors, particular care must be taken to make a gas-tight join between gasket strips, especially at the corners; this is best effected by liberal use of the adhesive material.

Door clamps: Door clamps are relied on to give a tight seal, and experience has shown that they should be placed not more than 30 cm (1 ft) apart all the way round the edge of the door. For overhead and hinged-type doors, refrigerator door fasteners are recommended. Small, hinged doors not greater than 1.8 m (20 ft^2) in area may be closed with ordinary sash fasteners 20 cm (8 in) apart. Sliding doors are forced against door frame gaskets by screw fasteners. If the floor of the chamber is above the outside floor or ground level, the clamps may be extended all along the bottom.

A sloping ramp is then required to load and unload the chamber. This may be drawn back when the door is closed. If the chamber floor is at outside level, it is not possible to clamp the bottom of the door unless special provision is made. A suggested arrangement is a shallow trench in the cement floor, which should be wide enough to allow sideways movements of the door and fasteners.

Circulation and venting: Proper circulation and post-fumigation venting of the fumigant/air mixture are essential in atmospheric chambers. Efficient circulation ensures that the fumigant is rapidly and evenly dispersed throughout the chamber so that no part of the load is overdosed or underdosed, while forced ventilation removes the fumigant so that the chamber can be safely unloaded after the treatment. Circulation and forced ventilation are essential in large fumigation chambers and advisable even for small ones.

A variety of fan systems can be used to achieve adequate circulation or ventilation and their size and capacity are governed by the volume of the chamber and the flow rate required. It is suggested that with most treatments of commodities the rate of air flow should give approximately one complete change of air every one to three minutes, based on the volume of the empty chamber. In selecting fans or blowers, due consideration must be given to friction in the exhaust pipe and other ducts in the system.

Suppliers of equipment will provide the necessary specifications of the equipment required for a particular installation. When nonflammable fumigants such as methyl bromide or ethylene dibromide are used, no protection from sparking is needed. With HCN, and other fumigants with wide flammability limits, the motors of fans should be totally enclosed or, preferably, placed outside the chamber, with the blade shaft inserted in a gas-tight connection in the wall. The system should be designed so that the same fan will be used for circulation and venting.

One method is to place the exhaust door, or port, near the fan so that when it is opened the fumigant/air mixture will be blown out into the open air. Another way is to use a blower with a large inlet and outlet. A diameter of 18 cm (7 in) is suitable. The outlet is fitted to a pipe of the same size. By means of a suitable blast gate or valve operated from outside, the air can be recirculated throughout the chamber by means of ducts or blown out through the exhaust stack into the open air.

If the ducts are used to draw the fumigant/air mixture from the bottom to the top of the chamber, distribution will be greatly improved from the beginning of the treatment. Ducts are best made from galvanised iron; the diameter required to give the suggested air flow of one complete change of air every one to three minutes will depend on the size of the chamber.

Exhaust door or port: The exhaust port may be a small door in the wall of the chamber; it should be at the opposite end of the chamber from the door so that after treatment, during venting and unloading, fresh air is drawn through the chamber from the open door. This exhaust port may slide open or move on hinges as a trapdoor. Experience has shown that the trapdoor may become a serious source of leakage. Great care must be taken to ensure that this door fits tightly in suitable gaskets and is firmly latched so that gas leakage is kept to a minimum. It is advisable to have the port in an accessible place so that it can be constantly checked.

Heating and lighting systems: A heating system is required if the chamber is used in cold weather. Sorption of fumigant on foodstuffs increases as the temperature decreases and, therefore, at lower temperatures larger amounts of fumigant are required and postfumigation ventilation is slower. As a general rule the temperature of the chamber and its load should be 15°C or above. If steam or hot water is available, it may be conducted through pipes set on opposite walls. A standard heating unit with blower may be used, which will also serve for fumigant circulation. However, such a blower should not be used for postfumigation venting unless the heat is turned off. Only nonflammable fumigants should be used in contact with steam pipes.

Electric or radiant heaters may be used if they provide adequate heat at short notice and if the elements are totally enclosed so that glowing wires may not come in contact with the fumigant. It is important to protect the heating units from mechanical damage and to protect the goods being fumigated from scorching or damage by the heater. Lighting inside the chamber is often necessary since windows are usually omitted from fumigation chambers. The lights should be arranged so that the loading does not obscure them and they must be adequately protected against damage.

Application equipment: Gaseous-type fumigants are introduced from outside the chamber through tubing. Copper tubing is preferred for permanent installations, although a type of plastic tubing, which is not affected by the fumigant, may be used. The size and type of tubing will vary according to the fumigant used. It is recommended that gaseous-type fumigants, such as HCN and methyl bromide propelled into the chamber under pressure, be discharged into a shallow galvanised iron evaporating pan, above 10 cm (4 in) wide and 5 cm (2 in) deep and two thirds the length of the chamber, which is suspended from the ceiling well above the load. This will speed evaporation and distribution, especially if the circulation fan blows downward above the surface of the liquid.

Liquid-type fumigants, such as ethylene dibromide, may be poured from a measuring cylinder into a small evaporating pan near the door. When nonflammable fumigants are used, they may be evaporated after the door is closed by warming the pan with a small, totally enclosed heater or heat lamp.

Accessories: There are several items of equipment that contribute to safety or help to achieve good results in the use of a fumigation chamber. All chambers that are not under constant surveillance during actual fumigation should be padlocked from outside. Also, a warning notice should be hung, or tacked, on the door while treatment is in progress. A small window or marine port permits a view of the inside of the chamber so that thermometers or other instruments can be read from the outside. One or more thermometers are essential for obtaining accurate readings of temperature in the free air space and in different parts of the load.

The thermometers may be entirely inside the chamber or they may have gauges or dials on the outside with cables leading to the sensitive bulbs placed, as required, inside the chamber. Sampling tubes for fumigant analysis, the number to be determined by the requirement to obtain samples of the gas from all representative parts of the space under fumigation, are useful accessories, which can aid in achieving efficient treatments. If they are installed at the time the chamber is built, they are available for use any time.

The most practical arrangement is to have several copper connections of about ().5 cm (0.25 in) outside diameter set at intervals at convenient points along one or more walls of the chamber. These are carefully sealed at both ends with removable caps or plugs. For sampling, the caps are removed and plastic or copper tubes leading to the gas analysis equipment are fitted to the outer ends of the desired sampling tubes. When a chamber is inside a building, it is advisable to have a red electric light bulb over the door. This is turned on during the treatment to indicate that gas is inside the chamber. With fixed chambers, it is desirable to have an offset control panel on which as many as possible of the recording instruments, electric switches, valve handles and other controls are situated. Such a unit simplifies operations and improves the appearance of the installation.

Pressure leakage test

Atmospheric chambers must retain the fumigant during the exposure period without appreciable loss through leakage to the surrounding atmosphere. The gas tightness of a chamber can be checked by a simple test where a positive air pressure is created and maintained in the chamber for a set length of time. An opening should be provided in the chamber to use a blower or other means to introduce air for creating the positive pressure. This pressure can be measured with an open-arm manometer filled with kerosene, which records the difference in kerosene levels in the two arms of the manometer.

The time lapse for the chamber pressure to recede from 50 to 5 mm should be 22 seconds or longer. Inability to develop or maintain adequate pressure indicates considerable leakage and the chamber should be checked for leaks at seams, gaskets and other points. Repairs should be continued until the time for the pressure to recede is more than 22 seconds. A smoke bomb or other device may be used in an effort to determine the areas of leakage.

Operation

Loading: The manner of loading the chamber will depend on the type of commodity and the method of handling it. A space of at least 30 cm (12 in) should be allowed between the top of the load and the ceiling. Goods can be stacked close to the side walls, provided they are not too near the heating units, but they should be kept clear of the end walls and at least 50 cm from the circulating fan. Bagged or packaged goods should not be placed directly on the floor. If these goods are not placed on platforms or pallets, a wooden floor rack with ample space between the slats should be provided. When methyl bromide is used, no extra space need be left between individual packages and bags because this fumigant penetrates well.

Temperature: The next step is to take the temperature of the material to be fumigated. Probe type "meat" thermometers have a pointed end and are suitable for placing in many materials. They are slow to respond, however, and sufficient time must be allowed for a correct reading. It is advisable to take temperatures in at least four well-separated points in the load. The temperature in the free air of the chamber should also be accurately determined.

For the purpose of selecting the appropriate dosage of fumigant, it is best to take the lowest temperature recorded, either in the commodity or the free space. As a general rule the temperature of the chamber and its load should be 15°C or above. Under certain circumstances, the fumigator should take into consideration the temperature at which the material was kept for several days before the treatment, because this factor might influence the response of the insects during the fumigation.

Computing the dosage: The dosages for various fumigants and commodities are given in the Schedules. In some instances the concentration x time (c x t) products have been worked out and are given also; their use is dependent upon an accurate determination of concentrations present in the chamber throughout the treatment. Table 1 gives the ml per 100 ft^3 (2.83 m^3), equivalent to certain dosages in terms of mg/l for a number of fumigants.

Closure: When it is time to start the fumigation, the door is firmly clamped. With most of the fumigants commonly used in chambers, the amounts introduced as dosage do not usually exert a significant positive pressure, especially when a full load sorbs some of the gas, or when there is a small amount of leakage. Therefore, all exhaust vents or ports must be tightly closed before the fumigant is introduced. If it is found in practice that considerable positive pressure is produced by the gas, one of the exhaust vents may be left slightly open while the fumigant is being introduced. If the fumigants that are used occupy a comparatively large part of the air space, 8 vent must be left open during application. Examples of such fumigants are the mixtures of ethylene oxide with carbon dioxide, and of ethylene dichloride with carbon tetrachloride.

Chambers used in the open may be equipped with small air pressure relief valves or exhaust tubes, which can be closed or capped as soon as application is finished.

Application of Fumigant

Gaseous-type fumigants

Fumigants, such as HCN and methyl bromide, are discharged from cylinders or other containers placed outside the chamber. When HCN is dispensed into an atmospheric chamber, it is necessary to apply additional pressure to the cylinder. This may be done with a tyre pump, in accordance with detailed instructions supplied by the manufacturer. When cylinders are used, they are placed on a platform scale and accurately weighed. The required dosage is then deducted from the total weight and the scale set at the lower point. The discharge valve on the cylinder is opened and closed quickly when the weight bar is again in balance.

Liquid-type fumigants

Fumigants which are liquids at room temperatures are poured into a shallow tray or trough inside the chamber or are poured onto burlap sacks or similar material from which they quickly evaporate. After the chamber door is closed, the circulating fans are started and the draught across the liquid or the sacked material hastens evaporation. As stated before, some fumigants with high boiling points, such as EDB, should be gently heated with a hot plate to hasten volatilisation.

Table 1. Dosage table for Fumigants used in smaller chambers

	lb/1 000 ft³							
	0.0625	*0.25*	*0.5*	*0.75*	*1*	*2*	*3*	*4*
	Millilitres of liquid/100 ft³							
Acrylonitrile 34% + carbon tetrachloride 66%	2.1	8.5	17.0	25.6	34.1	68.2	102.3	136.4
Carbon disulphide	2.2	9.0	17.9	26.9	35.9	71.7	107.6	143.5
Carbon tetrachloride	1.8	7.1	14.2	21.3	28.4	56.8	85.2	113.6
Chloropicrin	1.7	6.9	13.7	20.6	27.4	54.9	82.3	109.7
Ethylene cholorobromide	1.7	6.7	13.4	20.1	26.8	53.6	80.5	107.3
Ethylene dibromide	1.3	5.2	10.4	15.6	20.8	43.7,	62.6	87.4
Ethylene dichloride 75% + carbon tetrachloride 25%	2.1	8.4	16.9	25.3	33.8	67.5	101.3	135.1
Ethylene oxide at 7°C	3.2	12.7	25.5	38.2	51.0	102.0	153.0	204.0
Hydrogen cyanide	4.1	16.5	32.9	49.4	65.9	131.8	197.6	263.5
Methyl bromide at 0°C	1.6	6.5	13.0	19.5	26.1	52.2	78.3	104.4
Propylene oxide	3.4	13.6	27.3	40.9	54.5	109.1	163.6	218.2

HCN discs

If HCN discs are used in a chamber, they are scattered by the operator. In small chambers of up to 28 m³ (1 000 ft³) capacity, the discs can be scattered from the partly opened door. For this type of application, the fumigator must always wear a respirator.

Fan circulation

After the fumigant is applied in the closed chamber, the circulating fans are operated for 15 to 30 minutes. With many commodity treatments, this initial circulation will suffice.

Exposure time

The period of exposure to gaseous-type fumigants begins when the discharge of the fumigant is completed. With liquids and discs, the exposure should be timed from the moment the door is firmly closed. The exposure periods for the various kinds of treatment at atmospheric pressure are given in the fumigation schedules. Fruit, vegetables, plants, bulbs and nursery stock are exposed usually for 1.5 to 4 hours; seeds and plant products, for 16 to 24 hours. For particularly sorptive commodities that may substantially deplete fumigant concentration, some allowance for this depletion may be necessary.

Venting and aeration

At the end of the treatment, venting should be commenced by opening the exhaust port or valve and starting the fan. The chamber door should be opened slightly to allow fresh air to flow in. At least 10 or 15 minutes should elapse before the door is fully opened. The time of this interval will depend on a number of factors, but the door should not be fully opened until the operator is assured by appropriate chemical tests, instrumental tests, or from long experience, that it is safe to enter the chamber to begin unloading. First aid kits and gas masks should be available and in good condition at all times.

Mobile and portable chambers

Fumigation chambers which can be readily moved from one place to another provide a flexible means of disinfestation. The actual surface of water exposed to the fumigant is small and does not affect the gas concentration significantly.

Portable drum fumigator

A small fumigation chamber, about 200 1 (7 ft^3) in volume, can be easily made from a clean steel or iron drum of the same size (50 gallons), as described by Johnson. The best closure is a water seal arrangement with a lid fitting into a collar, or trough, built round the outside of the drum. The head of the drum is cut out with a cold chisel and the rough edges smoothed. The collar is made of 20- to 24- gauge galvanised sheet metal and is tightly soldered or welded to the drum.

The top of the collar should be about 2.5 cm (1 in) below the rim of the drum, so that if water is spilled it does not fall into the drum. A tap or plug near the bottom of the collar permits drainage. The lid or cover is made of the same material as the collar. It consists of a disc and a rim, which fit neatly into the collar. Disc and rim are carefully soldered or welded to provide a gas tight seam. One or two handles on the lid facilitate lifting. A small fan or blower is necessary for the proper distribution of the fumigant.

It may be operated by electricity from an ordinary lighting circuit or a battery, or manually by a hand-crank on the outside of the chamber with a chain drive connected to the blower axle. The manual arrangement is practical, as no reliance is placed on electric power. However, a special connexion must be fitted where the crank axle passes through the chamber wall. Heat may be provided by ordinary light bulbs, manually or thermostatically controlled. Fans and heating bulbs should be placed below the rack on which the load rests. This rack may be made of rat wire of 12 mm (0.5 in) mesh, or similar material.

Where continuous circulation is required in a small chamber, a suitable small fan, with a high impedance motor, can be connected in series with electric light bulbs located on the outside of the chamber to serve as a rheostat. In this manner, the fan may be operated continuously during the fumigation periods with no increase in temperature because of the heat developed by the motor. Equipment such as dispensing tubes, plungers and thermometers may be inserted through gas tight fittings, for example rubber stoppers, in holes in the lid or sides.

The water seal in the collar is satisfactory for fumigants such as methyl bromide or ethylene dibromide, which are only slightly soluble in water. Methyl bromide may be dispensed from the 280 ml applicator already described, or from 20 ml ampoules which are placed in clips in a tray inside the lid before it is closed; the ampoules are thus broken by the plunger. It should be pointed out that if an ampoule containing 20 ml of methyl bromide is released inside a drum of 200 l (approximately 50-gall) capacity, the resulting concentration will be about 170 mg/l (10.5 lb/1 000 ft^3 or 170 g/m^3). With this concentration, injury to growing plants may occur even if the exposure period is shortened to give an appropriate concentration x time (c x t) product.

If ethylene dibromide is used, the required amount is poured as a liquid into the evaporating pan. This can be done through a stopper after the lid is closed. Ethylene dibromide evaporates slowly and it should be heated with a small heating element without exposed wires, placed below the evaporating pan, in order that a full concentration of gas may be reached at the beginning of the treatment.

When the lid is placed in the water collar, it is necessary to release some of the air trapped inside the drum in order to allow the rim to be well below the water level and to prevent positive pressure being developed inside the drum. The air can be allowed to escape through one of the stoppers in the lid. The drum should be painted on the inside with a primer and then given a resinbased varnish or other finish appropriate to the fumigant principally used.

Mobile water-sealed fumigator

It is safe to use methyl bromide gas under an aluminium cover, but the liquid should never be left in contact with the metal for any length of time because of the possibility of fire. The air pressure release valve prevents the top section of the fumigator from being too buoyant and keeps it well seated in the water seal. Also, positive pressure created by the introduced fumigant may be released through this valve. Mobile fumigation chambers that have facilities for applying and circulating the fumigant and employ gaskets to give gas tight seals are produced commercially. Models with capacities of 10 and 20 m are available.

Plain steel drums with clamp-type lids

Where steel drums, equipped with removable, clamp-type covers and rubber gaskets, are available, they can be used as fumigation chambers without modification, except possibly for the holes required for the insertion of tubing and wire. The rubber gasket provides an adequate seal, but it should be checked occasionally for tightness.

5

Fumigation Under Sheets

Sheets retaining fumigant vapours for a sufficient time when placed over infested materials may provide a safe and effective method of fumigation. The word "tarpaulin" has in the past been used to describe these sheets, but this term is considered to be too general and may lead to a misunderstanding of the properties required for satisfactory treatments. Materials spitable for this type of work have been called "gas-proof sheets". This is a more descriptive term, although, in fact, all the materials so far used have been shown to differ somewhat in their permeability to fumigants. This technique has widened the usefulness of fumigants by making it possible to treat infested materials without moving them from their place of storage or diverting them from their usual trade channels.

Under some conditions it is possible to retain the sheets over the stacks of goods after fumigation and aeration, so that continuous protection is provided against reinfestation, contamination by bird droppings, water leaks, dust and dirt. In recent years, the development of new plastic materials has led to the introduction of types of sheeting which are well suited for fumigation covers. The sheets used for particular jobs vary according to the amount of rough handling and the weather likely to be encountered. There have been several critical studies on the diffusion of methyl bromide and phosphine through various materials which might be used for fumigation sheets. Sheets made of polyethylene or polyvinyl chloride film are suitable for use on the ground or to cover stacks in protected situations.

The types most often used for fumigation have a thickness of 0.1 mm (0.004 in). Sheets of this gauge weigh less than 100 9 per m (2 lb per 100 ft^2) and are easily handled even in large sizes. When sheets are to be used several times a thickness of 0.15 mm is recommended. Sheets made of these films tear easily on sharp corners or projections. They are not recommended for use on high buildings or structures when there is any possibility of high winds which may tear them to shreds.

In some countries where the sheeting material is inexpensive and easily obtainable, it is often more economical in terms of labour to discard low density sheets instead of folding them up for further use. For use in large-scale operations outdoors in exposed situations, coated fabrics are more suitable. Types that have been successfully used are nylon or terylene fabrics coated on troth sides with neoprene, polyvinyl chloride, or butyl rubber, and cotton fabrics, coated on both sides with neoprene. Coated cotton fabrics appear to be more permeable to methyl bromide than the others. It has been found that a smooth slippery surface makes handling easier, especially if both sides of the fabric are coated.

Manufacturers of coated nylon fumigation sheets supply them with ropes sewn in plain hems on all four sides to facilitate clamping large covers together. For most fumigation purposes, the sheets are purchased in rectangular shapes. In certain circumstances, the overlapping of sheets is inconvenient and in such cases tailored covers may be more practical. For example, for the fumigation of large pyramids of groundnuts in northern Nigeria special covers tailored to the shape of the pyramids are used.

The goods most often fumigated under sheets are cereals and other plant products in bags. Cereals in cartons, dried fruit in boxes and tobacco in boxes or hogsheads can also be treated in this way. The fumigation of buildings under sheets is described below.

An important consideration in this type of treatment is the most efficient utilisation of a sheet of a given area; the dimensions of the stack should be such as to provide the maximum volume. Today, the principal fumigants used for this work are methyl bromide and phosphine. Ethylene oxide and HCN have also been used, but HCN does not penetrate as well as methyl bromide.

In addition to the usual precautions taken before any treatment it is essential in fumigation under sheets that, after treatment, the gas escaping from the stack does not endanger any persons working or living in the vicinity. The fumigations may be conducted inside sheds, warehouses and similar structures, which furnish protection from bad weather. Such an arrangement is satisfactory if there is no possibility that gas may diffuse into occupied rooms or quarters.

Very often the work can be safely done in warehouse rooms in which the windows can be kept open. Suitable warning signs indicating type of fumigant and date of application should be placed on all stacks while fumigation is in progress, whether indoors or outdoors.

A first step is to make certain that there will be no leakage of fumigant downwards during treatment. There is no serious loss or danger if the stack is placed on firm ground or on a good cement floor. If the floor is not gasproof, another gas-proof sheet, or even rolls of tar paper overlapped and sealed with masking tape, may be placed on the floor and the stack built above it. When using methyl bromide or other similar fumigant it is essential that special provision be made to prevent the fumigant from discharging directly into sacks near the gas outlet. A suitable space may be provided by lifting and moving some bags from the top layer.

In small piles, four bags may be propped against each other to form a "dome". Under this is laid a pan or can to receive the liquid discharging from the end of the outlet tube. In larger stacks, a long space two bags deep and two bags wide should be made by lifting adjoining bags to form a trench. For these larger volumes, the outlets of the gas lines should be either spray nozzles, which ensure volatilisation of the fumigant before the liquid reaches the bags or the sheeting, or open tubes leading into pans or cans in the manner already described. If electrical power and outlets are available, one or more 25- or 30- cm (10 or 12 in) fans, the number varying according to the size of the stack, aid greatly in distributing the fumigant.

Sheets of standard sizes are easily placed on rectangular or square stacks. It is preferable to have sheets large enough to cover the individual stacks. At least one foot of sheeting should be left on all sides of the stack to provide room for sealing the edges to the floor. If the sheets are not large enough to cover the stack, one or more must be joined by rolling the edges together.

According to Brown, a satisfactory seal for methyl bromide fumigation may be obtained by rolling together about 1 m (3 ft) of the edges of each of the adjoining sheets. "The leading edge of the first sheet is folded back three feet. The second sheet is laid to overlap the first so that the edges lie together then these edges are held together and rolled until all the overlap is taken up. When covering a large stack, joins may have to be made in two directions." When sheets on stacks exposed to the wind or on buildings are joined, clamping of the rolled junctions is necessary.

Under some conditions, such as those encountered in west Africa with the large pyramidal stacks, the rolling together of sheets is impracticable. Sealing with sprayed plastic has been tried with success but the standard practice is to use tailored covers for these stacks. The edges of the sheets are best sealed to the floor by "snakes", plastic tubes about 10 cm (4 in) in diameter, filled with sand or water, which gives an efficient seal on level ground or floors. The use of water is very convenient because the tubes may be drained and rolled up between treatments.

In emergency, some of the bags of the material already fumigated may be used, but a seal is difficult to obtain with these. Chains weighing not less than 2 kg per m (1.5 lb per ft) may also be used, but care must be taken that these do not damage the fabric. Outdoors, earth may be dug and piled closely around the base of the stack. Rips and holes appearing in the fabric may be repaired with adhesive or masking tape.

When gaseous fumigants, especially those containing methyl bromide, are applied indoors, respirators must be worn. For large piles the fumigant is applied from cylinders under pressure. With cylinders of methyl bromide discharging into high stacks, additional pressure in the cylinders may be needed before the full dosage is dispensed. For many treatments with methyl bromide, 1 lb cans are convenient.

Aluminium Phosphide

Tablets, pellets, sachets or other formulations used for generating phosphine may be applied to the stack before it is covered with the sheets. With long stacks it may be safer or more convenient to apply them as the sheeting work proceeds. It is often convenient to place paper trays containing the proportional number of tablets or pellets in strategic positions throughout the stack. The sachets are obtainable in lots of ten, bound together by string. These strings may be draped over the stacks in evenly spaced positions before the fumigation sheets are laid on. Under normal circumstances, when the generation of the phosphine takes place slowly, it is not necessary to wear a respirator.

When tablets, pellets or sachets are directly handled, rubber gloves should be worn. After the edges of sheets are secured with sand snakes, tape or other suitable material, warning signs should be placed on all sides of the stack. If the stacks are properly sealed, workers need not vacate the premises but good crossventilation must be maintained during working hours and periodic tests with detector tubes or other suitable analyser should be made to ensure that no one is exposed to concentrations above the threshold limit value.

Fan Circulation

With gaseous fumigants, the fans should be operated for 15 minutes at the beginning of the treatment. Operation of the fans for a longer time will not greatly improve gas distribution and may tend to force the fumigant out, especially if the seal at the base of the sheets is poor. Fan circulation is not recommended for phosphine treatments and should not be necessary if proper distribution is made of the aluminium phosphide before fumigation begins.

Leakage

When methyl bromide is being used, a careful check should be made with the leak detector around the entire base of the stack immediately after the fan is turned off. If the covering is made of more than one sheet, the rolled junctions should also be tested. Any serious leak can thus be found and corrected at once.

Aeration

During the process of airing, the operators must wear respirators while working on the stack or in its vicinity, until it is shown by an appropriate test that dangerous concentrations of the fumigant are not present. Aeration should be begun by quickly lifting several sheets of each corner of the stack. After an interval, more of the sheeting may be lifted. When it has been demonstrated by the use of the leak detector, or by other chemical tests, that high concentrations of fumigant have diffused away, the sheets may be carefully drawn off the stack. Special precautions, such as opening all doors and windows and running exhaust fans, should be taken when airing a stack indoors. Careful checks should be made to determine that the indoor space is safe for human occupancy.

Paper Coverings

A specialised method of stack fumigation is the use of kraft paper for covering dried fruit. This technique is an inexpensive and simple means of effective fumigation and subsequent storage. It is particularly useful on farm. The enclosure is made by framing stacks of fruit boxes with 2.5 x 10 cm (1 x 4 in) timber and covering it with paper. The paper used by Barnes and Reilly consisted of two layers of heavy kraft paper with an asphalt laminant reinforced with sisal fibres.

The shape and construction of the stacks are important to success. They should be built on 10 x 10 cm (4 x 4 in) foundations to protect the fruit from ground moisture and to facilitate gas circulation. The frame is covered by carrying the paper up one side, across the top, and down the other side, allowance being made for 30 or 45 cm (12 or 18 in) widths for sealing on the ground. The paper strips are lapped about 8 cm (3 in) and then battened with laths. Since the cover papers shrink after weathering, allowance should be made for this by spacing the frame boards 2 m (about 7 ft) apart to take the edges of the 2.5 m (8 ft) wide paper.

The paper cover is sealed to the ground with sand or fine soil. The seal should cover the paper on the ground completely and should be 15 to 20 cm (6 to 8 in) high against the sides of the structure. Head room above the boxes is necessary for gas circulation; if the stacks are not under a roof, they should be peaked to allow rainwater to run off. Tears in the paper may be repaired with masking tape. After fumigation, several months' storage is provided by the enclosure, and reinfestation by insects is prevented. This technique, or modifications of it, could be applied to a number of similar storage problems.

Some buildings can be rendered gas tight only by the costly and time consuming sealing of individual cracks and holes, and it is often much easier to cover the entire structure with sheets. An early successful application of this technique was the commercial control of dry wood termites in dwellings. Formerly, thick kraft paper was laid over the buildings, but this material has been largely replaced by the modern types of plastic sheets already described. The effectiveness of this method of insect control was demonstrated during the campaign to eradicate the Khapra beetle in California, in the course of which many large mills, grain elevators and warehouses were completely covered and fumigated with methyl bromide.

In one of the treatments, the volume fumigated under covers was 113 000 cm (4 million ft^3). Rasmussen reported on a critical study of methyl bromide fumigation to control long-horn beetles in houses covered with gas-proof sheets during treatment. The technique for fumigating large structures is basically the same as that already described for stack fumigation. There are a number of special requirements, however, brought about mainly by the fact that the covering sheets may be situated high above the ground and be exposed to winds of considerable velocity.

Sheets: Although polyethylene sheeting may be used, the danger of its being torn by high winds, or even totally destroyed, has led to the use of the tougher, but much heavier, nylon or terylene fabrics coated with plastic. The weight of this material, best suited for covering buildings, varies between 200 and 440 g/m (6 and 13 oz/yd^2).

6

Fumigation in Large Buildings

Recently, phosphine generated from aluminium phosphide has been successfully applied for the fumigation of large buildings, such as warehouses, which can be rendered sufficiently gas tight to retain the vapours.

The structure to be fumigated should be carefully inspected at the outset so that unusual features or potential problems can be taken into consideration before the fumigation is started. Such matters as sources of leakage, commodities that will absorb large amounts of fumigant, possible damage to commodities, arrangements for gas analysis, safety of personnel, etc. should all be noted so that appropriate arrangements can be made beforehand. In preparation for fumigation, a check list of the various duties to be performed should be made—a fumigator should never trust to memory or assume that essential things have been done. It is important to instruct the fumigation crew carefully, show them any special problems or hazards and inform them of emergency procedures in case of an accident.

The use of fumigants is often regulated by local or state governments. Regulations are designed to protect the health of those applying the chemicals and of the public in general. Regardless of whether the treatments are made regularly in permanent installations or are carried out once only in a particular place, it is necessary that all regulations be observed. It is usually advisable to inform local fire departments of large-scale treatments, especially if an entire structure or building is to be fumigated. The police should also be warned if the work is being done in a built-up area.

If the fumigation is being done in a place accessible to the public or adjoining public thoroughfares, it is necessary to employ one or more guards to keep unauthorised persons from approaching the fumigation site or from interfering with the operation. The guards are employed to keep outsiders away while the fumigation is being prepared, during the gas application, during the treatment and until the entire fumigation area has been declared "gas free" after the treatment has ended.

Regardless of whether guards are employed or not, properly worded warning signs should be posted on all sides of a fumigation site. These signs should have clearly printed letters at least 5 cm (2 in) high, with wording such as:

- Danger
- Fumigation with poison gas
- Keep out by order

The date of application of the fumigant should be included on the warning notices for those treatments where there is a possibility that the notices may not be removed at the end of the fumigation, e.g. on freight containers, railway box cars or other transportation vehicles. Notices printed on cardboard are often available from the fumigant manufacturer or dealer.

Fresh, unused respirator canisters, specifically designed for protection against the particular fumigant to be employed, should be provided for each operator likely to be exposed to any concentration of the gas. A minimum number is two canisters per operator. One is to be used during application of the fumigant and, if there is exposure to any fumigant during this operation, it should be replaced by a new one before the aeration process begins. Careful checks of individual respirators, carried out under the supervision of the foreman or leader, should be made by all operators concerned.

Fumigation Preparations

Sealing

Particular attention must be given to the proper sealing of a structure, as this may constitute the difference between the success and failure of a fumigation. Doors and windows should be firmly closed and almost invariably should be taped or caulked.

Circulating Fans

When methyl bromide is being used, artificial circulation should be provided by means of fans to prevent stratification of this heavy gas and to ensure even distribution. One 40 cm (16 in) fan is usually sufficient for every 1 400 m^3 (50 000 ft^3). Provision should be made to switch the fans on or off from outside the building.

Plan for Gas Discharge

All fumigations which are not strictly routine in nature must be carefully planned. If a particular operation is complicated or conducted on a large scale, it is well to have the duties and movements of each employee typed out on a specially drawn-up work sheet. When fumigating buildings, warehouses or ships with more than one floor or deck, it is advisable to include a plan on this sheet showing the location of each fumigant cylinder and gas outlet and all exits and entrances. When crews are fumigating large structures by releasing the fumigant from inside, it is sometimes advisable to place directional signs at strategic stairways, doors or turnings.

Sometimes it is helpful to chalk arrows and other marks on the floor to indicate the direction and order of movements. The work should be organised so that at no time will the operators pass through a room or space in which the gas has already been discharged. In buildings without basements, discharge should start on the top floor in order that the crew may work toward the prearranged exit on the ground floor. If there is a basement or cellar, a special plan may have to be made.

Usually, the main floor may be omitted while one crew releases the fumigant in the upper floors and another works in the basement. The main floor is not treated until the others have been traversed and all operators, except those concerned with the final discharge, have left the building. It is most important that operators should not be releasing fumigants inside a building for more than 30 minutes. It is better that a 15- to 20- minute period be allowed for each person. If the gas release cannot be done properly within that time by the available staff, more workers should be allocated to the crew during the preliminary planning.

In large operations, it is good practice to have the crew working in pairs; if an accident occurs to one person, the other can bring him into the fresh air or summon assistance. Each member should be fully acquainted with the role of the others in the team. When the discharge of the fumigant is to be done by operators moving inside the structure, a full rehearsal of the procedure should be carried out at least once before the actual fumigation is started.

Dosage Recommendations

All dosage recommendations are made on the basis of the volume of the structure. For regularly shaped square or rectangular buildings, dosages are calculated by multiplying length by width by height. If the buildings are irregular, the volume of each unit should be calculated separately, and then all added together. If the building has a peaked roof, the average between the height of the sidewall and the distance from the lowest floor to the top of the roof may be taken as the height dimension in calculating the volume of the structure. No allowance should be made for space occupied by commodities, materials, machinery or furnishings.

Provision for Ventilation

If possible, some provision should be made in advance for starting postfumigation ventilation before the fumigators enter the buildings. If front and rear doors can be opened from the outside, a strong draught can be created to dilute the gas concentrations appreciably when ventilation is started. When the roof ventilators are covered with plastic sheeting or kraft paper, they can be quickly and safely uncovered at ventilation time and the fans can be started.

When all preparations are completed, several checks should be made before the fumigants are applied.

1. *Sources of ignition*: All possible sources of fire should be eliminated. All pilot lights and gas and oil burners should be extinguished. It is particularly

important that thermostatically controlled, high temperature electrical equipment of any kind be disconnected during the fumigation. HCN is inflammable at the high concentrations that may be present during initial discharge of the gas. Methyl bromide may react in the presence of a flame or glowing wire to form hydrobromic acid, which may be corrosive or injurious to many materials.

2. *Warning signs*: Ascertain that the warning signs are properly placed on all possible entrances to the building, and, in large-scale operations, at strategic points on the approaches to the building.
3. *Outside telephone*: Locate a nearby telephone of ready and constant access which may be used in case of emergency.
4. *Guards*: See that the guards or watchmen are at their proper stations before gas discharge begins. They should prevent unauthorised persons from entering the building or interfering with the work of the fumigation crew.
5. *Check on evacuation*: As a final essential measure, the foreman fumigator, accompanied if necessary by senior crew members, should visit every floor and room of the structure calling in a loud voice a warning, such as: "Poison gas fumigation to begin everybody out." Failure to observe this precaution in the past has led to serious accidents, including at least one fatality. During the visit these operators will also have an opportunity to see if all other preparations are properly completed.

Application of the Fumigant

While the fumigant is being applied, all persons engaged in or associated with this operation should wear respirators. The only permissible exception to this rule concerns the operators working in the open, or in some well-ventilated place, under conditions in which any gas that escapes from the equipment is immediately diluted and dissipated. The respirators should not be removed until the workers indoors have reached fresh air, the fumigant has been completely discharged, and all the valves and piping have been closed so that no fumigant can escape from the system. During the application, unauthorised persons should not be allowed to approach or talk to the operators engaged in the discharge of the fumigant.

As an aid to effective fumigation, sampling tubes should be connected to strategic points throughout the structure. During the fumigation, samples may be drawn at intervals and analysed by one of the methods already described for the fumigant being used. If concentrations at any point fall below an established level, more fumigant may be introduced as required.

Ventilation

The operators should wear respirators when they approach or enter the building.

In accordance with preparations made before fumigation, when airing begins, as many doors and windows as possible are opened from the outside. Ventilators and fans with readily accessible switches may be started. The operators should then withdraw from the immediate vicinity of the fumigated structure. At least 30 minutes should elapse before the fumigators enter the building to open more doors and windows and to start other ventilators.

At the beginning of aeration, the building should be entered only for short periods of time. The process of ventilation may be divided into stages, with ample time given to each one. After being in the building for a short period, the operators should withdraw into fresh air, remove their respirators, and wait for 15 minutes or more before putting them on again and re-entering the structure.

As soon as the crew foreman determines that the structure is properly aired, the appropriate chemical tests for residual fumigant, already described, may be carried out. Particular attention should be paid to the retention of gas in highly sorptive material such as flours, meals and jute bags. This retention may be unduly prolonged by introducing cold air into a building that was warm during the fumigation.

Serious accidents may result from closing such a building again and warming it up and causing the release of sorbed gas into the atmosphere. In cold weather, therefore, after the preliminary aeration, all doors and windows should be closed, the building heated to above 24°C for two or three hours, and the aeration repeated. Under exceptional conditions, this process may have to be repeated two or three times before all material is safely free of gas. After chemical tests have shown that both the structure and its contents are free from toxic concentrations, the building may be declared gas free and reoccupation permitted.

Application of HCN

The generation of HCN by the action of sulphuric acid on a cyanide salt is usually referred to as the pot method, although wooden barrels are often used as well as crocks or pots. This is not as convenient as the other two methods described below, which have largely replaced it, but it lends itself to emergency treatments.

Sulphuric acid, 66° Beaume, is usually supplied in 45 1 (10 gall) carboys in wooden frames. It is advisable to divide this into smaller amounts in bottles; for this operation, a tilting frame is recommended. Acid should be poured slowly and with caution. The sodium cyanide eggs, usually 25 9 (1 oz) each, should be kept in tightly closed containers, away from heat and moisture, until they are needed.

Sodium cyanide is a powerful stomach poison, which can also cause serious poisoning by being absorbed through cuts and exposed skin. All persons handling the dry sodium cyanide, and mixing the sulphuric acid in water before fumigation begins, should wear rubber gloves and tight fitting goggles. Chemicals are made ready in the following proportions according to the systems of measurement used.

	Metric	*British*	*U.S.*
Sodium cyanide	1.0 kg	1.0 lb	1.0 lb
Sulphuric acid	1.5 litres	1.2 pints	1.5 pints
Water	3.0 litres	2.4 pints	3.0 pints

If properly generated according to the method given here, 0.45 kg (1 lb) of sodium cyanide will generate 0.25 kg (8 oz approximately) of HCN gas. For many building fumigations at temperatures above 20°C, the dosage of HCN gas recommended is 8 g/m^3 (8oz/1 000 ft^3).

Use and Placing of Generators

Earthenware crocks up to 18 1 (4 gall) capacity or clean, water-tight 227 1 (50 gall) wooden barrels should be used as generators. For large-scale operations in mills and warehouses, the barrels are safer to use. The generators may be set out in groups of three or four on each floor of the building, each in a galvanised iron washtub, if available. The tub should contain a pailful of water with several handfuls of washing soda to neutralise any acid that may leak or spill from the barrel.

After the generators are set out in their final positions, the required amount of water is poured into each. Then the required amount of acid for each generator is measured into pails. Next, the acid is poured carefully and slowly into the water in the generators.

The next step is to count the number of sodium eggs required for each generator and to place them in double-walled kraft paper bags or flour sacks. The bags or sacks are then carefully placed 30 or 60 cm (1 or 2 ft) away from the generator, where they can be easily seen and reached.

Generating the Fumigant

After the final preparation and inspection of the buildings have been made, the respirators, fitted with anticyanide canisters, are put on. Gas generation is started at the farthest point from the final exit. As each generator is reached, the operator carefully lowers the bag with cyanide into the acid, without allowing any liquid to splash on his person. Then he moves quickly to the next generator and repeats the same procedure at each station until he reaches the final exit.

Cleaning Generators

After the fumigation is finished and the structure has been thoroughly ventilated, the residue must be emptied and the generators cleaned. Sometimes the chemical reaction is incomplete and therefore it is advisable to wear a respirator while handling the generators and emptying them. The residual material should be taken to properly approved sites for disposal. The empty generators should then be thoroughly washed with fresh water and the liquid disposed of in the manner already suggested.

Special Pressure Generators

Special generators made of acid-resistant metals are available from some manufacturers of sodium cyanide. They permit the generating of HCN outside a structure. The acid is placed at the bottom of the metal generator and the sodium cyanide, preferably in egg-shaped form, is suspended above in a metal basket. The lid of the generator is then tightly clamped down with thumbscrews. When the charge of HCN is to be released, the basket of sodium cyanide eggs is plunged into the acid and the generated gas flows through a rubber hose into the fumigation space. This method is especially suitable for releasing HCN into fixed installations.

Application of Liquid HCN

If liquid HCN is available in steel cylinders, this is one of the most convenient ways of applying it on a large scale. This method is commonly used in countries where HCN is manufactured, or where it may readily be shipped. The cylinders are always discharged from the outside of the structures being fumigated. This feature adds to the convenience and safety of the method. For fumigation of buildings and ships, the HCN must be forced under pressure through piping. For this purpose, air compressors are connected to a special valve on the head of the cylinder to drive the liquid through a siphon tube out of the discharge valve into the lines.

Some premises, which are regularly fumigated with HCN, have a permanent piping system of copper tubes and the necessary connections. When buildings are fumigated occasionally, a temporary piping system made of rubber hose is used. The use of HCN absorbed in porous materials such as cardboard discs, which has already been described, is a convenient method of dispensing the gas. The hermetically sealed cans hold a standard amount of HCN, usually 0.5 or 1 kg (1 or 2.25 lb) approximately. Some brands are available with small amounts of added warning agents, such as chloropicrin. The use of HCN discs must be carefully planned according to the system already described.

Cooling: The discs should be precooled for several hours before treatment begins. This can be done by immersing the cans in tubs of water containing ice, by placing them in a refrigerator overnight, or by surrounding them with dry ice. The use of dry ice is especially suitable during hot weather.

Protection of floors: To prevent floors or floor coverings from being marred, the discs may be emptied onto sheets of corrugated or ordinary wrapping paper of about 1 m^2 (1 yd).

Opening cans: The cans are quickly opened with the special can opener. Respirators must always be worn when cans of discs are opened or distributed.

Distributing cans: In the fumigation of large buildings with two or more floors or many rooms, it is advisable to open the cans at the points of distribution and immediately cap them with the fibre caps that act as protection for the cans during shipment. When fumigation is started, this provision enables the operator to empty the cans and scatter the discs.

Scattering the discs: The fumigating crew of not less than two men should work according to a prearranged plan and proceed toward the final exit in the manner already described. In hot weather, HCN evolves rapidly from the discs and the operators may have to move quickly to avoid working in poisonous atmospheres.

Disposing of discs: Under normal conditions, the used discs are free of HCN by the time the structure is opened for ventilation. Nevertheless, all discs should be carefully gathered and burned in the open, or taken to an approved disposal site.

Application of Liquid Methyl Bromide

Liquid methyl bromide should not come directly in contact with painted surfaces. This can be avoided by volatilising the spray in the free air or by discharging the liquid into metal pans or cons with a large surface area.

The fans provided for methyl bromide fumigation should be operated for 15 to 10 minutes at the beginning of a fumigation. Once equal distribution has been attained, the gas/air mixture remains in equilibrium almost indefinitely and the methyl bromide does not stratify during the period of fumigation. However, if gas analysis at any time indicates unequal distribution, or any additional fumigant is discharged during treatment, the fans should he operated again for as long as is considered necessary.

Methyl bromide cylinders may be placed, according to dosage requirements at strategic points throughout the structure so that the valves may be fully opened by the operators, wearing respirators, as they retreat toward the final exit. In large operations, or in structures with high ceilings or roofs, stand pipes (curved copper tubing directed upward) are often attached to the cylinder outlets. In this technique, a short "T" is fitted to the top of the pipe to discharge the gas laterally and prevent contact with the ceiling.

Hot gas method

Methyl bromide may be discharged from outside the building by passing it through a heat exchanger after it leaves the cylinder. This is the method already described for the fumigation of buildings under tarpaulins.

Recirculation method

Another method, applicable in low structures that are not covered by tarpaulins, is the use of plastic ducts similar to those used in the fumigation of grain by the recirculation technique. To prevent the collapse of plastic ducts at bends, short metal ducts of the same diameter are introduced into the line wherever it is necessary to change direction. A fan draws air from the ground floor and forces it through the duct back into the building through an upstairs window. Methyl bromide is introduced through tubing on the intake side of the fan and is thus distributed throughout the building.

Pressure method

For discharging methyl bromide into flour mills or other tall buildings from the outside, the fumigant may he forced through the lines by compressed air. The liquid flows through flexible copper tubing of 9 or 12 mm (0.375 or 0.5 in) outside diameter. At the outlet end, each tube is pinched to give an aperture of 2 mm (0.08 in) width. The outlets are directed laterally or upward, but never downward. Some manufacturers add nitrogen or another inert gas above the liquid in the cylinder in order to supplement the natural vapour pressure of the fumigant. This combination is usually sufficient to start the methyl bromide through the lines, even in tall buildings. When the cylinder is half empty, additional pressure up to 10.5 kg/cm^2 (50 lb/in^2) may be provided from a compressor in order to complete the discharge.

Application of Phosphine

A number of formulations for generating phosphine are available for the fumigation of large structures. When tablets or pellets of aluminium phosphide are used, they are distributed on trays or thick paper at intervals throughout the building. Similarly, magnesium phosphide plates are laid out to give an even dispersion of the required dosage. If there are one or more storeys, the operators start at the top floor and work downward toward the final exit. Under normal conditions evolution of the phosphine is delayed and it is not necessary to wear respirators if the distribution can be completed in a few minutes.

If there is a possibility of considerable delay, or under conditions of high temperature and humidity, respirators should be worn at the discretion of the official supervising the fumigation. For small structures, where the work can be completed in a few minutes, sachets may be employed by suspending them evenly on thin rope as from a washing line.

Mill Fumigation

Methyl bromide and phosphine are effective in the fumigation of modern concrete, stone or brick mills that can be made sufficiently gas tight to retain the fumigant. HCN may be satisfactory for mills which allow more leakage, such as ones with frame or sheet metal construction number of special measures must be carried out when mills are fumigated in order that the fumigant may penetrate all parts of the machinery and equipment. These are especially important with HCN because it does not penetrate well into accumulations of flour, meal or other sorptive materials.

HCN Fumigation

The following is a summary of the more important measures to be taken:

1. *Before stopping the mill*: Shut off the source of grain fed to the mill. Then continue to run the machinery until 811 parts of the milling system are empty. At the same time, remove elevator bootslides so that stock may be pulled out and not carried forward. Use a rubber mallet or a similar device that will not damage the equipment to hammer all parts of the machinery from which accumulations of flour, meal or dust can be dislodged.
2. *After stopping the mill*: All machines, ducts, bins and conveyors in the entire mill should be thoroughly cleaned. All infested material collected in cleaning the mill should be removed to the fumigation chamber and fumigated or the debris should be taken outdoors and buried. All doors to conveyor boxes and bins, and those leading to enclosed machinery, should be opened as widely as possible to aid gas penetration. It is especially important to open the "dead" spouts into which the fumigant penetrates with difficulty.
3. *Machinery piping system*: Fumigation of the various units of the mill and its machinery is assisted by the installation of a special piping system equipped with nozzles which automatically deliver equal quantities of fumigant to each unit. An installation of this type leads to a considerable saving of fumigant. Also, it may eliminate the necessity of dismantling and cleaning the machinery, and labour costs may thus be greatly reduced. The manufacturers of liquid HCN supply information on the planning and installation of such systems.

Methyl Bromide and Phosphine Fumigation

These two fumigants penetrate well into accumulations of sorptive mill products and trash lying in a mill. Therefore, if methyl bromide is used, the cleaning programme required before HCN fumigation can be delayed until after the fumigation is completed. In fact, this is recommended because the insects are killed where they are and the debris does not need to be moved to a special fumigation chamber. All doors or vents to bins, conveyors and machinery should be opened as already outlined for HCN fumigation.

Local Fumigation

Local fumigation is the treatment of isolated insect infestations in buildings. In mills, the term is applied particularly to the fumigation of individual machines and enclosed equipment that harbour insects. The popular term "spot fumigation" applies to this type of treatment as well as to treatments of isolated outbreaks in stored grain. The fumigants used for this work are pure compounds or mixtures that evaporate slowly and do not diffuse quickly from the treated locality into the main space of the building. Although these vapours diffuse slowly, they kill insects only when the equipment is fairly tightly enclosed. It is usual to apply the fumigants during weekends or at night after the mill has been shut down. They should be applied regularly every two or three weeks.

A regular programme of local fumigations may keep insect populations at a low level throughout the year. This contributes to the production of goods free from infestation. However, only in exceptional circumstances does such a programme eliminate the need for occasional general fumigations. Local fumigation may be done by alert and competent mill employees who are assigned to this work and specially trained for it.

Before local fumigants are applied, the mill should be run dry. The milling stock should not be removed until after treatment. All vents from milling machinery to the outside should be closed and other steps taken to reduce draughts in the mill system as much as possible. Dead spouts and filled suction lines should be cleaned. In all applications of local fumigants, the operators should wear industrial type respirators fitted with the proper canisters. Care should be taken not to spill liquids on skin or clothing.

Application Methods

There are two principal methods for the application of local fumigants in mills: hand application and injection by automatic applicators.

Hand Application

When the fumigants are applied by hand, the dosages for each locality may be poured from the shipping drums into litre or quart bottles, which are then placed individually beside each particular point of application. It is best to use bottles graduated in millilitres or fluid ounces so that the correct dosage may be applied at each station. The operators should work in pairs. With hand applications, the usual procedure is to start at the top of the building and work quickly downward from floor to floor as the liquids are poured or dashed into the individual units.

After each application, the opening is tightly closed. The empty bottles may be abandoned and picked up when the building is finally aerated after fumigation. Windows should be left open during application but closed on each floor as the team works downward. When application is completed, all doors or other entrances should be closed and locked. Warning signs should be posted at all possible entrances and a guard placed on duty; these and other precautions should be generally the same as for all building fumigations.

If possible, the mill should not be opened until 24 hours have elapsed from the time of application. If only one night i allowed for treatment, exposure may be shortened to 12 or 16 hours. At the end of exposure, all doors and windows should be opened to provide as much ventilation as possible. Appropriate tests for lingering fumigant should be made with detector tubes or other suitable equipment to ensure that gas concentrations are below the threshold limit value before the building is reoccupied.

Automatic Application

This type of application is confined to treatments of milling machinery. The fumigant is discharged in equal, predetermined dosages from special equipment operated by compressed air, which are supplied by the fumigant manufacturers or blenders. The liquid is injected through small holes specially bored at selected points in the milling machinery. Because this is essentially a machinery fumigation, the building may be fully ventilated during treatment.

Automatic injection in mills is a specialised treatment and details of dosage, location of injection points and other technical considerations are determined after an inspection of the premises has been made by representatives of the companies supplying the fumigants and the applicators. A similar type of treatment has been used for the control of insects in the wooden linings of railway box cars.

7

Fumigation in Vehicles

Fumigation of Bagged Goods in Ships and Barges

Plant products in bags may readily be fumigated with methyl bromide while they are still loaded in ships and barges. If proper provision is made for the even distribution of the fumigant throughout the load any full cargo space may be fumigated. An important consideration is that in large spaces it may take some time to complete thorough aeration after treatment. A procedure for the use of phosphine in combination with methyl bromide for the fumigation of bulk loaded expeller in barges and coasting vessels has been developed by Wohlgemuth et al.

Methyl Bromide Fumigation

The methyl bromide cylinders are placed on the open deck and connected to copper or polyethylene tubing leading into the holds and ending in one or more T-shaped nozzles. At least one 25- or 30-cm (10- or 12-in) circulating fan should be placed as near as possible to the bottom of each hold and, when the gas is released, the fans should be operated for 30 minutes in order to prevent stratification and to aid distribution of the yes. Fans are also useful in hastening aeration after fumigation. The penetration of methyl bromide is outstanding and the gas will escape unless adequate precautions are taken for sealing.

A single ordinary canvas tarpaulin is not sufficient to contain the gas during the fumigation period. However, three canvas tarpaulins in good condition are able to prevent leakage, especially if they are dampened. Suitable sizes of special tarpaulins with a plastic coating impervious to methyl bromide may be obtained to cover the hatches. If suitable or sufficient tarpaulins are not available, strips of tar paper joined together by various widths of brown paper glued with flour paste should be laid down above the hatches, with ordinary canvas tarpaulins then placed on top.

Canvas tarpaulins usually provide sufficient cover for the ventilators. Many modern ships have steel rolling hatch covers, called "McGregor type", which require no extra sealing for methyl bromide fumigation. These covers are lowered into place off their dollies onto a rubber gasket and the full weight is sufficient for a good seal. If the dollies are warped after rough use the side bolts may be used to pull them down into place. When methyl bromide is used, the ships' officers, with their knowledge of the construction of the vessel, may be of great assistance to ensure that adequate sealing is obtained. They can point out possible sources of leakage, help check the ventilation system and openings between bulkhead on older ships and also see that hatch covers are secured properly.

Phosphine Fumigation

Phosphine may be applied in barges loaded with bagged goods. If the cargo space of a vessel is fumigated while the ship is in port, it is recommended that the entire crew be evacuated while the fumigation is in progress. Warning notices should be prominently displayed at gangways and entrances to accommodations and a guard should be stationed at each gangway to prevent unauthorised persons from going aboard. No one should be allowed to return to quarters until the ship has been certified "gas free" by the fumigator in charge.

Small-scale Treatments

On board ship there may be rooms usable as fumigation chambers. Ammunition and other storage lockers are usually gas-tight and of sufficient size for treating small amounts of infested goods. Fumigation is conducted in the same way as in a regular chamber. Extra care must be taken during aeration because, in the absence of a proper exhaust vent, the gas will diffuse out through doors and portholes.

Fumigation of Empty Cargo Spaces

Empty holds in cargo ships or on some passenger vessels often have residual populations of stored-product insects. If the infestation is localised, it can be controlled by spraying with a suitable insecticide. This method, however, cannot be relied upon for complete eradication. Usually, a complete fumigation is required.

Aerosols, which are fogs, mists or insecticidal smokes produced by various devices, have already been mentioned in the Introduction. They can be used to treat populations of stored-product insects in empty holds, and are satisfactory for killing the insects exposed in the open. However, all of these materials, including dichlorvos, have poor powers of penetration and are not as efficient as fumigants for obtaining complete control throughout the structure.

When a fumigant is used, one or two boards covering the bilges in each side of the ship should be removed. Piles of dunnage, if present, should be broken up to allow penetration of the gas from every side.

Application of Methyl Bromide

Methyl bromide penetrates well into cracks, crevices and spaces under boards. In such places there are accumulations of grain and other foods in which the insect populations are able to multiply and from which they spread to the cargoes and other parts of the ship. The fumigation of empty holds with methyl bromide follows the same basic procedure already described for loaded cargo spaces.

Fans placed at the bottom of the holds are again absolutely necessary to give proper gas distribution. It will be noted that dosages are lower and exposure periods shorter than those for loaded spaces. To eradicate Irogoderma beetle infestations, more particularly Irogoderma qranarium Everts, increased dosages of methyl bromide are needed.

Application of Hydrogen Cyanide

HCN may also be used in empty holds although it does not diffuse very well through residual debris from plant products to reach deep-seated infestations. The most convenient method to apply HCN in ships is by means of the impregnated discs. These are often kept on hand for medical quarantine fumigation against rats. With HCN, the general preparations and precautions are the same as for methyl bromide, with the following points of difference.

1. The ordinary canvas hatch tarpaulins, if in good condition, are satisfactory for HCN fumigation; it is advisable to use two to each hatch.

2. If fans are used for distribution of HCN within the hold, they must be explosion proof.
3. The discs are distributed as widely as possible in the hold by lifting the tarpaulins on each corner of the hatch in turn; this work is facilitated by removing the four corner hatch boards before the fumigation.
4. The tarpaulins are then battened down securely.

Aeration

Procedure and precautions for the ventilation of empty cargo ships are the same as already described for loaded ships. The gases diffuse more quickly from the empty holds and, under conditions favourable for aeration, the holds may be entered two or three hours after the hatches are opened.

Fumigation in Wheeled Carriers

Some wheeled carriers may be used for fumigation. Treatment in such vehicles is often convenient and may show considerable economy in time and labour. By making a treatment of infested goods in a carrier, at least four manipulations, involving loading and unloading the vehicle and the fumigation chamber, are avoided. Furthermore, the fumigation kills the insects in the free space of the carrier and live pests do not remain behind to infest or invade the next load. In this way a great deal of possible cross-infestation may be avoided.

Railway Cars

Railway companies are usually willing to permit fumigations in their equipment. It is, of course, essential to solicit their cooperation before treatments are planned. In some countries, the standard railway cars (wagons) are well built and constitute in themselves excellent fumigation chambers; steel cars are especially suitable. Refrigerator-type cars, by the very nature of their construction, are usually gas tight if all openings are carefully sealed. In all railway cars, the wooden floor is likely to be the principal site of leakage.

Some cars that are not sufficiently gas tight may be readily sealed by some of the methods already described. Also, 8 leaky car could be covered with a gas-proof sheet and fumigated, if a satisfactory seal can be made at ground level and around the rails. When railway cars are to be fumigated, they are usually isolated on separate sidings and kept there while under gas. This arrangement is, of course, essential if the cars are covered with sheets. In some countries, under certain circumstances, railway cars may be allowed to travel while still under gas.

Appropriate warning signs are placed on the main doors or hatches, with instructions for the adequate airing of the car before it is entered for inspection or unloading. In France and in the United States, there are special fumigation chambers with railway tracks running into them so that the cars may be fumigated singly or in groups. Some of the chambers are designed for fumigation at atmospheric and others at reduced pressures. This arrangement has the advantage of allowing treatment of both the inside and outside of the vehicles, a consideration which may be important from the quarantine point of view, especially at border points.

The fumigation of railway cars under gas-proof sheets may seem to eliminate the need of constructing permanent chambers. In plant quarantine practice, however, the choice of method against the threat of introduced pests should be made on the grounds of efficacy rather than of convenience or economy.

Trucks and Cargo Trailers

Highway vehicles that can be rendered gas tight are suitable for the fumigating of goods loaded into them. In general, the same considerations apply as for railway cars. For certain operations, the use of gas-proof sheets, particularly of the light-weight polyethylene type, is a convenient and effective improvisation. In this work, the trucks or trailers may be driven onto a gasproof groundsheet and then covered with another sheet, which is sealed to the lower one with snakes in the usual manner, thus providing a gasproof structure around the vehicle.

Freight Containers

Fumigation of many commodities is carried out in freight containers and other transport units that are designed for carriage on vehicles. The procedures for treatment of these containers are much the same as for railway cars and cargo trailers. They may be fumigated while stationary or the treatment may be continued "in transit". Procedures for fumigation of tobacco with phosphine in freight containers, along with efficacy of the treatment, have been outlined by Childs et al.

Application of Fumigants

In practice, methyl bromide and phosphine are the materials most commmonly used for wheeled carriers containing plant products, because these two yeses penetrate effectively into many commodities. Ethylene dibromide was used successfully in the United States for truck fumigation in the campaign against the Mediterranean fruit fly in Florida. Truckloads of oranges were run into atmospheric chambers equipped with ducts specially designed for circulating the fumigant/air mixture through the load.

Methyl bromide

Fumigation of bagged grains, meals and other plant products in carriers, such as railway cars, is easily done with methyl bromide in cans (0.45 or 0.68 kg) or in steel cylinders. This fumigant should always be applied from outside a railway car using 6 mm (0.25 in) polyethylene or copper tubing attached to the cylinder or to the special opener/applicator designed for cans. The tubing is inserted through a crack between the door and door frame, through a roof vent or possibly through a hole drilled in the floor.

The discharge end of the tube is plugged sod a hole drilled through both walls 3-4 cm below the tip, then the end of the tube is attached close to the ceiling at the centre of the car, so that the methyl bromide mist can be directed over the commodity and toward both ends of the car. The dosage to be applied is determined according to the size of boxcar as well as the temperatures of the atmosphere and the commodity. Temperatures above 10°C with exposure periods of 12 to 18 hours are recommended for this treatment.

After applying the proper dosage the tubing is withdrawn and the opening sealed. At the end of the fumigation period all doors and vents are opened to allow as much circulation as possible. At least 30 to 60 minutes are required to aerate the free space of the car, but appropriate detecting equipment should be used to make sure that no methyl bromide is present when personnel enter to unload the car. Respiratory protective equipment should be worn when checking for fumigant in the car.

Phosphine

Various formulations for generating phosphine are used

in fumigating railway cars and other carriers. Tablets or pellets of aluminium phosphide may be used by placing them in moisture-permeable envelopes and attaching these at intervals to the wall of the carrier or placing them in shallow cardboard boxes on top of the load. To avoid the possibility of combustion, only two tablets or ten pellets should be inserted in each envelope. Prepacks, with pellets appropriately separated from one another in a specially prepared plastic strip and covered with permeable paper, are supplied by some manufacturers for convenient application of the fumigant. They ore sometimes attached to the inner side of the door just prior to closing and sealing.

If sachets are used they may be strung by a thin rope above the load. Lochner gave a full description of the use of aluminium phosphide tablets for the fumigation of maize in bags or in bulk while in transit in railway cars. Magnesium phosphide in tablets, pellets or embedded in moisture-permeable plastic plates is also available for use in a similar way.

Safety Measures

If a line of cars is being treated, it is advisable to put the usual gas warning sign on a special stand, which is staked in a conspicuous spot beside or between the railway tracks at each end of the line. Railway sidings chosen for holding cars during fumigation should be at least 100 m from any dwelling or building regularly occupied. Where in-transit treatments are carried out, warning signs should be placed in conspicuous locations near each door of the carrier. They should be firmly fixed so that they are not easily lost or removed before the car or container is opened for aeration.

For such in-transit treatments the warning signs should indicate the date of fumigation as well as the type of fumigant. Prior to unloading of fumigated vehicles, appropriate tests should be made to ascertain safety of the cargo area. The success of in-transit fumigations is particularly dependent on good construction of the railway car or freight container and effective sealing methods.

Fan Circulation

Fan circulation is essential in carrier treatments for fresh fruit, vegetables, plants and nursery stock when using a fumigant such as methyl bromide. This is to ensure proper evaporation and distribution of the fumigant in treatments of short duration (2 to 4 hours). In fumigation of bagged grain and plant products, the exposure periods are at least 12 hours. If the fumigant is evenly distributed above the load at the time of application, subsequent fan circulation is unnecessary.

Fumigations of Bulk Grain

Fumigations of bulk grain may be carried out in railway cars and in wellbuilt trucks. The principles and techniques are essentially the same as for bulk grains in flat storages. Aluminium phosphide tablets applied by probes or sachets inserted in the bulk may be used. Liquidtype fumigants have also been used in open-top transit trucks with the grain being covered with a gas-proof sheet after application. A canvas tarpaulin can be used as an additional cover. Gaseous-type fumigants such as methyl bromide, may also be used if provision is made for adequate recirculation with special equipment.

Fumigation in Individual Package

There are at present two factors which may assist the development of such methods for practical use. First, some modern packaging materials are sufficiently gas tight to retain the vapours for the length of time required for successful treatment. Second, there are fumigants which can be applied conveniently to containers and which also have physical properties, such as slow evaporation or diffusion, that are suitable for this purpose.

In India, ethylene dibromide was injected into grain, stored in jute bags, at the rate of 10 ml per bag. This either killed the insects present or drove them out. Experiments with injection of four other fumigants into insecticide-impregnated jute bags containing grain confirmed the greater efficacy of EDB for this purpose. It should be noted that EDB is now considered to be hazardous to human health and, therefore, care should be taken to ensure that no one is exposed to its vapours and treater grain should be thoroughly aired so that residue is reduced below the FAD guideline levels.

In the United Kingdom, two tablets of aluminium phosphide were inserted individually into bags of wheat and rye resting on gas-proof sheets. One tablet was placed near the middle anti one near the mouth of the hag. Tablets were deposited either by hand or with a special probe dispenser. The gasproof sheets were then wrapped over small groups of bays and well overlapped.

After five days of exposure, the sheets were removed. This treatment gave good control of a representative collection of species of stored-product insects and mites, although some adults of the granary weevil, Sitophilus granaries, survived in cold wheat used in one test. It appears that this method holds promise for treatments of sacked grain in farm storages or other places where little special skill is required for the control of certain insect species, such as the granary weevil, the dosage per bag should be increased to three tablets.

Using polyethylene-lined sacks, Proctor and Ashman achieved good control of insects in bags of groundnuts being transported from Zambia to the U.K.. When a 0.6 9 aluminium phosphide pellet was applied in jute or woven polypropylene sacks of 32-82 kg capacity and lined with polyethylene 63.5127.5 μm thick, the c x t product for phosphine exceeded 50 mg h/l, and 10°C percent mortality of all stages of the test insects, including Sitophilus zeamais Mots was recorded. Although many liners in sacks split during transportation, this did not affect the efficiency of the phosphine fumigation or permit reinfestation.

Packaging Line Treatments

Treatments on the packaging line are of concern at present in the food processing industries, and the subject falls more in the realm of industrial rather than agricultural practice. Packaging line treatments have helped to keep certain food products, such as dried fruit and vegetables, free from infestation as they leave the processing plant.

The technique is applicable when the packages are made of materials, such as cellophane and fibreboard, through which fumigants do not diffuse very quickly. The fumigants are dispensed into the individual packages from automatic machines, which can be calibrated to give an accurate dose. Application may be made before or after the contents are placed in the package. The best stage is just before the package enters the unit which seals the wrapping. The wrappings generally used are sufficiently gas tight to maintain an insecticidal concentration of the fumigant long enough to kill any insect stages present in the package.

To protect plant workers, exhaust hoods, approved by health authorities, should be installed to draw fumes away from the working area. Also, appropriate analytical tests should be made to ensure that threshold limit values for the fumigants are not exceeded. Experience has shown that fumigants diffuse away from the packages fairly rapidly after application, sometimes within 48 hours. Thus, it is considered that these treatments do not create any hazard to consumers.

Fumigants used for this type of application include methyl bromide, ethyl acetate, ethyl formate and propylene oxide. In the past, methyl bromide has been mixed with high proportions of carbon tetrachloride for package treatments. Propylene oxide and ethylene oxide are also applied to certain packaged foods, especially dried fruit, to prevent microbial spoilage and to control insects. Dosages for packaging line treatments vary greatly according to the product, material used for packing, method of packing and fumigant used.

8

Fumigation for the Control of Insect Pest in Storage

Fumigation as a control technique may be described simply as the establishment of an athmosphere containing a lethal gas in the environment of an insect of a concentration high enough and an exposure period long enough to kill the insect.

Factors Affecting the Toxicity of Fumigants

Temperature

Temperature greatly affects the toxicity of fumigants to insect. At a temperature below 15°C, stored product insects are not as active as at higher temperatures and tend to absorb less fumigant over a given period of time. Chemical reactivity and rates of diffusion of fumigants are also reduced by lowering the temperature. From 10°C to 35°C, the concentration of a fumigation required to kill a given stage of an insect species decreases with the rise in temperature.

Adsorption/Desorption

Adsorption is the most important physical factor modifying the penetration of a fumigant. Physical sorption of the fumigant by the material containing the insect is reduced when the temperature increases and proportionately highed concentrations of fumigant are needed to kill the insects. As the temperature is lowered, the amount of gas physically adsorbed increases and it is necessary to add more fumigant in order to sustain concentrations free to act on the insects.

Table 1. Required concentration x time (cf) products to obtain 99% mortality of Tenebroides mauritanicus

Concentration methyl bromide	Exposure hours	Ct product g hr/m
83	2	166
55.3	3	166
41.5	4	166
33.2	5	166
23.7	7	166
16.6	10	166

Also, at low temperatures, diffusion of the gas into porous commodities is slower and there is a corresponding decrease in the rate of desorption afterwards. Chemical reaction of the fumigant with some of the fumigated materials increases as the temperature is raised. If the residue formed is significant, it is advisable to conduct the treatment at as low temperature as permissible.

Relative Humidity

Variation in response at certain humidities have been observed not only between different species subjected to different fumigants, but also between stages of the same species exposed to a single fumigant.

Carbon Dioxide

Cotton et al Jones. and Bond et al have demonstrated that the addition of CO_2 to some of the fumigants may increase or accolerate the toxic effect of the gas by simulating the respiratory movements and opening of the spiracle of the insect. However, excessive amount may tend to anaesthesize the insects thereby enterfering the action of the fumigants.

Narcosis

At high concentrations phosphine acts as a narcotic and the tolerance of insect is greatly increased. In adults of a susceptible strain of T. castaneum a seventy fold increase in tolerance at the LD99 level occurs when they are exposed to a concentration of 50 mg/ compared with that of 0.5 mg/l. Although narcosis is symptomatic of a high concentration it does not appear to be the neehanesm to resistance.

Fumigation Enclosures

The choice of fumigant and its method of application are largely determined by the situation and type of enclosures in which the treatment is to be undertaken. Fumigation enclosures for grain may be classified as follows:

1. Sealed fixed enclosures
   - fumigation chambers
   - silo bins
   - storage sheds or godown
2. Sealed movable enclosures
   - road and rail vehicle
   - barges and ship

- freight container

3. Temporary enclosures
   - under gas proof sheet
   - spot fumigation

Fumigation of Bagged Grain

Proper storage practices permit good fumigation and help maintain a level of sanitation that prolongs the time between fumigations. They help pin-point the sources of infestation and try to eliminate sources of insect breeding. Proper sanitation requires proper placement of stacks so that sweeping, inspection, and pesticide treatment can take place. The following should be observed:

- Stacks should be no more than 6 meters high
- Minimum gap between stacks and wall or columns should be 1 meter.
- Minimum gap between stacks for gang ways should be 2 meters.

Before bags are stored in any area, it should be cleaned of all debris and loose grain. All sweepings should be fumigated or immediately destroyed to avoid harbouring an infestation. Floor cracks should be filled with cement or hard filler. Floors need spraying with a good residual pesticide to discourage insect migration for storage.

Floors should dry before any material is placed in the area. Walls should be checked for holes or cracks that could retain old grain. Moth webbing and old grain should be brushed clean and all holes and cracks should be filled with cement. Wall spraying with residual pesticides, of or appropriate type and formulation could be considered when practical to do so. Overhead areas should be cleaned of dust and moth webbing and sprayed with residual insecticides. Storage should be segregated:

- Old and new stock
- Good and bad stock
- Different moisture levels
- Different grains or varieties

Incoming grain should be checked for insect infestation and treated if infested. A regular programme of inspection maintenance and sanitation should be implemented.

Sanitation Requirements for Stored Bagged Grain

- All grain should be inspected on arrival and fumigated if infested.
- Insects will live in spilled grain next to bags and along walls. This must be swept up daily and fumigated or destroyed.
- Insects can live in grain dust on ledges and ceiling beams. These must be kept clean enough to discourage this infestation.
- Tailings and siftings from equipment will contain many live insects. This should be fumigated or destroyed each day.
- Weekly inspections should be made to locate possible infestations or grain that is no longer in good condition.
- Where market demands permit, the oldest stock should always be shipped first.
- Applications of residual insecticides should be made to the walls and other insect resting areas at least once per month.

- If the storage area can be sealed enough for effective treatment, space treatments with Dichlorovos should be made twice per month.
- All used bags should be fumigated before reuse.
- All cleaning and bagging equipment should be fumigated once per month minimum.
- Stock that can no longer be sold at regular prices and is more of an insect attractant than a value, should be disposed of at the least possible loss. Destruction may still be more profitable than keeping the material and increasing treatment costs.
- Broken bags should be repaired and spills cleaned up as soon as found.
- A rodent trapping or poisoning programme should be maintained wherever rodents are a problem.
- Bird infestation in the godowns should be discouraged through use of hanging strips at doorways and other mechanical deterrents.
- Floor drains should be kept clean and treated to discourage cockroach infestations which can affect the odor and flavour of rice.

Code of Practice for Fumigation Techniques

Preparation of area to be fumigated

- Check to be sure there is enough room on all sides and at the top to work. If bags are too close to walls or posts, the fumigation should not be attempted until it can be done safely and effectively.
- Have the area around the stack swept clean so that a good floor seal can be achieved.
- Measure the length width and height of the stack and calculate the volume to be fumigated.

- Inspect to determine the type of insects causing infestation.
- Calculate the dosage based on volume and species to be controlled
- Consider potential reinfestation and include nearby stacks in fumigation if infested.
- Consider safety of potentially exposed workers. Will there be other workers in this area during the fumigation? Can the fumigation be done at another time? Is there enough cross ventilation to provide a level of safety? Is gas monitoring equipment available if workers must be nearby? All safety feretions should be very safety taken into consider workers reservation.
- Choose the proper fumigants.
- Choose Methyl bromide if:
  a) Less than 4 days are available
  b) There is a heavy infestation of grain mites
  c) Nearby electrical equipment could be damaged
  d) Odor of chloropicrin is needed for safety reasons.
  e) The price is advantageous
- Choose Phosphine if:
  a) Four or more days are available
  b) Deep penetration is needed
  c) Product may have high level of bromine from previous fumigations.
  d) Damage to germination is a concern
  e) The price is advantageous

- Be sure that all needles equipment is on hand for sealing closing, leakage detection, safety equipment, ventilation

Equipment needed for fumigation of bagged grain

- Measuring tape to determine volume
- Broom to clean floor
- Ladder if stack is hard to climb
- Polymer vinyl sheets or other gas proof sheets
- Mending tape for sheets
- Warning signs
- Respirators
- Leak detectors
- Instruments to measure concentration
- Gloves if required
- Adequate amount of fumigant
- Label and labelling for fumigant or other instructions.
- Clamps for rolled joints
- Sandsnakes
- Thermometer
- Trays if phosphine is used
- Scales if methyl bromide is used Fans if methyl bromide is used
- Hoses if methyl bromide is used.

Proper placement of gas-proof sheets

Most gas-proof sheets can be easily torn and must be handled with a degree of care. The first step is to prefold the sheets so that they can be unrolled or unfolded easily.

Many fumigators prefer to lay the sheet flat in a large open area and have one man on each side of the short direction of the sheet. Working together, they pick up an edge of the sheet about 1/4 of the way from the middle and bring it to the mid-point. Other sections are folded on top in accordian style until there is a stack on either side of the midpoint. The two sides are folded on top of each other and then one end of the folded area that is now about 1/2 meter 1 meter wide is pulled on top for a distance of about 2/3 of the length of the lower section.

The sheet is then rolled tightly working from the folded end to the open ends so that air will be worked out. This tightly rolled sheet is then easy to transport and to lift on top of stacks. The rolled sheet is placed on top of the stack and unrolled along the longest direction of the stack. The end of the folded sheet is pulled down so that is touches the floor and overlaps by at least 1.0 m. One side is then, pulled down until it too touches the floor and overlaps by 1 m. If this sheet is even on the floor it will facilitate other work. Sand snakes can then be used to anchor this area.

Next, pull the other edge down until it too touches the floor with a 1 m. overlap. If there is surplus sheet, it is better to gather it on the top in folds and hold it there with sand snakes. Surplus sheet on the floor makes sealing more difficult. The corners are the hardest part to seal correctly since if they are not done corners, there will be too many folds on the floor and gas may leak. The corner is in effect a large triangle of material. Some try to roll this triangle and this has advantages outside where there are strong winds. However inside, the rolled corners will always result in gas loss at the floor.

The best way is to anchor one edge of the corner with a sand snake and then fold the surplus back along the other edge with an inside fold. This can be anchored with a sand snake and the sheet will then fold back down with two folded corners coming together at the corner. Along the floor there will be only large smooth sections that can be sealed with the sand snakes. The major leaking point will be at the floor level snakes.

The sand snakes should be pieced in a double row so that the second row will overlap the joints between sand snakes in the first row. At the corners, sand snakes should go around the corner rather than meet at the corner. Floor level leakage also occurs when there is a floor joint, a crack, or a floor drain. These areas should be avoided when preparing for a fumigation but the problem can be partially Overcome by taping over these openings. If a single gas proof sheet will not cover the stack, it will be necessary to join two or more sheets together.

Just overlapping the sheets will not retain the gas for long enough. The sheets can be joined together and be reasonably gas tight by taking one meter of the edge of each sheet and rolling the two together into a tight seam. This rolled joint is then clamped with steel spring clamps every 1 meter or less. The rolled joint requires extra care for sealing at the bottom. The seam should be flattened and extra sand snakes used to get a better seal. This is still not a strong seam. If it was placed horizontally the weight of the sheet would pull the seam apart after a few hours.

The seams must run vertically or be gathered on top of the stack and weighted with sand snakes so that there is no pull against the seam. If "Hanging seam" must be used such as when tall buildings are covered, another procedure is used. The two sheet edges are placed between two long boards, the combination is then rolled together several times and retained with C clamps every 1-2 meters.

If two sheets are not enough and only a small amount of additional sheet is needed, it should be placed in the middle. It is important of fold the middle sheet so that the combined sheets fit evenly and tightly. Lay the middle sheet on the floor and fold it in half. If the gap to be filled is 4 meters plus a meter on each side to roll, then we want a 6 meter sheet. Roll the folded side of the sheet until within 3 meters of the edge. This will open up to be a 6 meter wide sheet with a false seam down the centre. Clamp the centre false seam and then place the middle sheet in place and join with the other two sheets. Clamps on rolled joints help prevent leakage Of gas during fumigation.

When many sheets are joined together, there could be a problem where two vertical seams meet at a horizontal seam. The seams must be staggered as shown in the illustration to avoid this.

Suggested dosage per bagged grain under gas proof sheets over 25°C

Phoshine

4 Days 5g/ton

5g/ton = 5g/1.6 m^3 Maize = 3.125 g/m^3

5g/ton = 5g/1.9 m^3 Paddy = 2.631 g/m^3

7 Days 2g/tonne

2g/ton = 29/1.6 m^3 maize = 1.25g/m^3 = 35.49/1000 cu. ft. maize

2g/ton = 2g/1.9 m^3 paddy = 1.05 g/m^3

Methyl bromide

24 Hours

9 lbs/100 ton = 28.8g/m³ = 1.81b/1000cu. = 28.8 oz/1000 cu. ft.

9 lbs/100 ton = 9 lb/5000 cu. ft. = 1.8 lb/100 cu. ft. = 140 OZ/62.5 m³

Conversion factors

g/m³ phosphine x 658.8 = ppm

ppm3 ÷ 10,000 = % by volume in air

g/m methyl bromide x 235.9 = ppm

oz/1000 cu.ft. = g/m = mg/liter (approximately)

Ib/1000 U9 bu x 12.87 = g/m

1 g/m³ phosphine = 659 ppm3 = 28.59/1000 cu. ft. 339/1165. n =.0077 m³/sec

1 ft/hr = 28 m³/hr =.47m³/min =.0077 m³/sec

1 picul = 60 kg

1 quintal = 100 kg

2.5 rai = 1 acre

Application of phosphine

After the cubic content and/or the weight of the material to be fumigated is determined, the chart wil: give the rate in grams per unit. Various forms of phosphine will yield different amount of gas;

1 tablet = 1 gram

5 pellets = 1 gram 33 9 bags = 11 grams

1 mg PH3 Plate = 33 grams

The success of the fumigation depends on maintaining a sufficient concentration for a long enough period of time. Aluminum phosphide releases gas slower than magnesium phosphide and bags of aluminum phosphide release gas slower than tablets which in turn release gas slower than pellets. Proper phosphine fumigations are for more than four days. It may be advantageous to use a combination of slow and fast release formulations or, if it can be done safely, to add part of the dosage after two days in order to prolong the period of moderate concentration. When tablets or pellets are used in fumigation of bagged grain, they should be placed in a container so that the final dust can be disposed of safely.

The containers must be large enough so that tablets are not piled on top of each other. Piled tablets are capable of sealing themselves so that the concentration in the centre of the stack can reach the point of combustion. When spread out in a container this is not a problem. Bags can be hung in a series on the side of a stack to help distribute the gas. Plates should be left standing on end on the floor with both sides open as they lean against the stack of bags or the pallets.

Gloves should be worn even though phosphine cannot be sorbed through the skin. There will be small amount of extremely toxic phosphide on the hands which could be ingested when eating or smoking after the fumigation if hand washing is not thorough. Even though phosphine formulations are not supposed to normally give off significant amount of gas in the first hour, it is safer to cover the stack completely before the phosphine produce materials are opened and applied. Release depends on temperature, moisture, and formulations.

Magnesium phosphide formulations have been measured to give of over 20 ppm in less than 30 minutes in mill fumigations. The manufacturer now suggests respiratory protection for any concentration over .1 ppm. Respiratory protection should either be worn or readily available and in good working order depending on the like hood of its need.

After the gas releasing material has been put in place and gas-proof sheets replaced, warning signs should be placed on all sides of the area where someone might approach. A final check for leaks should then be made. The following description of termination of the phosphine fumigation presumes that the sheets are required for another fumigation. If they are not required, and if the grain is thoroughly dry, the sheets may be left in position. If the stack has been a long time under sheets it may be safe to assume that no gas remains.

1. Place a respirator around the neck.
2. Open all doors and windows.
3. Check the phosphine concentration using detector tubes and put on the respirator if the concentration is unsafe. If more than 0.1 ppm, send unprotected workers to another area
4. Remove the sand snakes, raise the corner of the sheets, the rolled joints on the vertical sides and quickly leave the area.
5. Leave for 30 minutes to one hour.
6. Return to open the sheet still further and again leave for a similar period.
7. Check on the safety of the area with detector tubes.
8. Remove the sheets, take away and fold for the next fumigation.
9. After a period of at least one hour, but preferably overnight, remove the spent residues and dispose of them in the recommended manner (Set "Disposal of Spent Formulation")
10. Remove the warning notices.
11. Complete a written record of the fumigation in a log book and attach a record to the stack.

Residues of pellets or tablets may be scattered on waste ground if amounts are small otherwise, they should be stirred into buckets containing warm water and detergent (30-60g of detergent in 10-25) of water. When no further bubbles can be seen in the liquid, the buckets should be emptied on waste ground. Containers should be immersed in the liquid and then buried. Bags of aluminium phosphide should be buried. A gas mask should be worn or care taken not to inhale the fumes during disposal. Residues from tablets, pellets and bags must not be heaped together, otherwise there is a risk of fire explosion.

Application of methyl bromide fumigator

- Over the stack with sheets, join up, fold the corners and place the sand snakes in position.
- Cheek sheeting, sand snakes and repair sheets with plastic tape if necessary.
- Inert hose through tarp into expansion chamber.
- Make final thorough check of the area around for possible hazards.
- Send away all onlookers from the fumigation area.
- Light halide lamps and put respirators on. Rubber gloves should not be worn.
- Record the weight of the cylinder. Or select the proper number of cans.
- Open the cylinder valve and check for any leaks. Introduce a small quantity of methyl bromide, about 10 percent of the full dose, by opening the side valve for a short period.
- Check and seal off any leaks around the stack. The halide lamp will indicate whether there is any need for a respirator to be worn.

– Re-open the side valve and introduce the methyl bromide until the scales indicate that the full dose has been given. Or add remaining cans of methyl bromide

Fumigation for containers

The success of fumigations depends on many small details being done correctly. Safety of the fumigator, other possibly exposed persons and cargo need to be considered. It is not enough to put the correct dosage into a container and shut the door. If the gas does not reach the insects in all parts of the container, they will not be killed. Gas can fail to reach the insects either through leakage or through poor circulation particularly with methyl bromide.

Examine the container carefully for any possible leaks around doors, hatches, or walls. These must be sealed before the fumigation is complete. Check the commodity and note whether suggested fumigant is appropriate. For example methyl bromide will not penetrate plastic bags. Chloropicrin is not legal on processed food going to the United States. Phosphine should not be used on starch vegetables like potatoes.

Phosphine could also hurt copper, gold or silver but these are not likely to be in a container that needed to be fumigated. If the gas is inappropriate, the shipper should be contacted. If methyl bromide is to be used, the tube should be taped to the roof of the interior of the container so that the gas will be discharged into the open space at the top where it will mix with the air before filtering down through the cargo. The hose can extend out through a corner of a door and the door carefully closed without normally pinching the hose.

Check the listed size of the container. Normally the cu. cap. is the most important figure since the entire space will be fumigated regardless of how full the container is. Calculate the dosage from the cubic capacity a volume, the net weight of commodity or from a chart considering both items. Put the required amount of gas next to each container. Several forms of phosphine can be used.

Grain that has not been processed such as bulk maize could be probed wish tablets. Rice in bulk or in bags could be fumigated with Phostoxin prepac, Detia Bags FumiCel plates or similar brands. Plates should not be dropped into processed foods such as polished rice since the plates may break apart during release. They can be put in cotton bags and suspended from the top. Plates should never be placed between bags or stacked on top of each other as a fire could result.

Methyl bromide is often used froms 1 1/2 lb cans instead of cylinders because of the small dose. If the gas is released into the head space it will probably be effective but the fumigator can be injured utilising these small containers. There is some leakage at the puncture point any time that the can is used. The leaks are worse if the seam on the can is punctured or if the gasket is old. Proper precautions are needed.

- Do not wear gloves, rings, or bandages on the hands
- Protect the eyes with at least goggles.
- Puncture can quickly but carefully being sure to miss any seams.
- Change the rubber gasket regularly.
- Set the can down and stand back while it is discharging to avoid skin burns.

- Pick can up for last discharge and hold away from body. Keep it high but tilted down to release last amount.
- Take clamp off the can when all gas seems to be discharged but keep in mind that there is probably a little bit left and can may "spit" Place empty can to the side until fumigation is done. Do not step on it as this could force gas out onto the foot.
- Proceed with next can or if this is the last, pull tube out through the door very carefully and drain the hose by holding it up. Many injuries occur at this point from the small amount of gas left in the hose being flipped onto someone.
- If methyl bromide gets on shoes or clothing, remove them immediately regardless of possible embarassment and do not put clothes back on till they have been washed. An extra set of clothes should be carried in the truck. Shoes may require a month of aeration.
- Fumigant cans and packages should be aerated thoroughly and then put in approved waste containers. They should not be carried inside the truck until thoroughly aerated.
- Warning signs should be placed at least on two sides to warn the crew not to store fumigated containers below the deck.
- Racks should be built on the roof or the sides of the truck for hoses that have been used for fumigation. They should not be carried inside a closed truck.
- Check seal around entire container with leak detector and seal any leaks before leaving
- Record work and gas used.

Fumigation for river barges

Phosphine tablets: Phosphine has the best chance of giving a good kill of all insects in bulk grain, but needs at least 4 days and a very good seal to maintain a lethal concentration. Even longer exposures may be necessary for resistant or tolerant species, or life stages. Ideally the phosphine producing tablets should be added as the grain is placed in the barge if mechanical equipment is available so that the barge can be loaded in only a few hours. In most cases tablets will have to be probed into the grain mass after the barge is loaded.

If a four day exposure is contemplated, a dosage of 5 9. per tonne is suggested. The expensive fumigant can be reduced to only 3 9. per tonne if a 7 day exposure can be used. For a four day exposure, it is best to probe the tablets to 3/4 of the depth of the hold. Longer exposures can permit shorter probes which can be done with less labour. In a typical barge a minimum of 30 places would need to be probed.

About 100 tablets would be used per probe but these should added a few each time the probe is withdrawn a few feet. Proper sealing is very important. Any openings from the hold to the rope locker or other areas should be sealed on both the hold side before loading and on the other side to minimise the chance of leakage. No one should be on the barge during the fumigation except possibly to take concentration readings. After the fumigation, the traps can be pulled back and the area should be kept secure from people until the concentration is below 13 ppm.

FumiCel and Fumistrips: The FumiCel and FumiStrip formulations of magnesium phosphide are a convenient new way of applying phosphine gas. These like the magnesium phosphide tablets release the gas faster than the older aluminum phosphide tablets but the kill many not be any quicker. Some insects must be exposed to the gas for a period of time and undergo at least one growth stage or molt before they succumb to the gas.

The FumiStrips can be used more effectively if they are fastened to the metal overhead rods so that they will be suspended in the air space instead of just being placed on the surface. They must never be used where two plates are touching each other. The FumiCel plates release 33 g. of phosphine gas and the strips release 528 g. of phosphine gas. A combination of the two can permit right dosage and again 5 9. per tonne is suggested for a four day exposure.

The manufacturer suggest that this be a static fumigation with the barge at anchor during this time. As with any fumigation, the quality of seal is very important. The type of fabric used in Thailand can hold phosphine very well if the tarp is new or is adequately patched. It is important for the fumigator to carefully examine the tarp and patch any obvious holes and it is important that they also monitor for leaks so that they can spot small hard to see holes or tears.

Compartment doors from the hold are best sealed with a section of plastic placed over the doorway. The door can then be shut on the sheet and the edges taped to the bulkhead. It can also be taped directly to the bulkhead over the door frame if there would be too many wrinkles in the tarp which could allow gas to escape. Warning signs must be posted at all potential entrance points to the barge and no one should be on the barge during fumigation unless readings show the air to contain less than 0.3 ppm.

Aeration with the tarps off should continue until gas readings are below 0.3 ppm even in the bottom of the hold and in the rope locker and other areas where the gas might linger. On warm windy days this should occur within a few hours but measurements are the safest way of knowing that the area is safe.

Fumigation on ship

Fumigations are done either at anchor or unnecessary because of previous infestation in a poorly cleaned hold but is needed more often due to infestation in the grain that is loaded. Fumigations are done either at anchor or intransit. The high cost of demurrage when ships are held at anchor makes the in-transit fumigation attractive but there are safety problems

Small leaks of gas can contaminate crews quarters, paint lockers and other areas and have resulted in some deaths. Intransit fumigation should only be done if a competent person has examined the ship and certified that a fumigation can be done without undue risk. A trained person then must accompany the ship and take gas readings throughout the fumigation so that any leakage will be discovered and appropriate actions can be taken.

Fumigation of bulk grain is very difficult because it is difficult to get even penetration of the gas. It is almost impossible to get good penetration of bulk grain with methyl bromide and penetration with phosphine can take from 7-14 days depending on the depth of the hold. In some tests conducted by USDA, adequate gas penetration was not achieved in the bottom of the hold unless ducts were installed prior to loading. Fumigation in-transit is a valuable technique since there is no later point to provide reinfestation.

Methyl bromide

Although methyl bromide does not have the penetraling

and dispersing power of phosphine, it can kill insects in 24 hours if the gas reaches all areas. Obviously the problem with methyl bromide is due to the fact that it is 3.3 times the weight of air and moves very slowly if there are no fans to circulate the gas.

If the gas is released at only one point at the centre of the barge as often happens, there will probably be failure to kill all of the insects. If the gas is released on top of the cargo at many different points, it will slowly drift down through the carge and above better distribution. Fans are still advised for recirculation of methyl bromide on barges.

On some ship fumigations, methyl bromide is dispersed through tubing with several notches cut along the length. Unfortunately most of the gas will be lost at the first few openings and only a little gas will come out of the last openings. A "H" configuration as shown below can solve this problem. The main hose is attached to a 3 meter section of hose with a brass "T". Then perforated hoses are attached at the centre with more brass "T"s. About four of these harnesses would be needed for the average barge and if they are made from a good quality plastic, they could be used many times. This would require releasing the gas into four hoses at four points instead of just one spot but this would not take much longer.

The dispersion of the gas will be greatly improved if the harness is supported above the bags by fastening it to the metal curved ribs of the barge. The use of a fan under the tarp for at least one hour after the introduction of the gas will greatly increase the dispersion of the gas. It is inexpensive to check for gas leaks of methyl bromide with a halide leak detector, and this should be done on each treatment so that additional sealing can be done.

Since there can still be small undiscovered leaks, no one should be on the barge during the fumigation unless equipped with a gas mask. Signs should be placed to warn all persons to stay away during the fumigation period. After 24 hours, the tarps can be removed and aeration can start but gas is likely to linger in lower areas and these areas should be checked before workers enter for unloading.

Ship fumigation while evacuated by personnel and crew

If there is a possibility that gas can leak from the holds to an occupied or potentially occupied area, fumigations should only be done after the crew and all other persons have left the ship. The only persons remaining on the ship should be fumigation personnel that are equipped with respiratory protection and gas measurement devices.

Before the gas is introduced, the fumigator in charge should plan how the gas will be aerated to a safe level. This may mean placing fans in some areas if electricity is available or can be made available with portable generators. If two days or less are available, only methyl bromide should be used since phosphine could not give an adequate kill in a short time.

Bulk grain cannot be adequately fumigated with methyl bromide unless perforated pipes are inserted deep into the grain in many areas to aid to distribution. This is a hazardous procedure that should only be attempted by trained personnel wearing adequate respiratory protection equipment. Bulk grain should normally only be fumigated with phosphine and with an exposure time of at least 4 days.

Bagged grain can be easily fumigated with methyl bromide if a series of release points are used such as the "H" harness discussed in the barge fumigation section.

Good control is still dependent on air circulation to in sure the gas stays evenly dispersed and can be readly evacuated. Fans must be used for at least one hour after gas release. This may mean gaining power through a portable generator. After the gas been released at many points and the fans have run for at least one hour, the fans can be shut off and the seal can be checked for leaks. All leaks should be patched by persons wearing respiratory equipment.

Exposure time should be at least 24 hours. After the exposure time is achieved, it is helpful if a new gas reading is taken to know if excessive leakage has occured. Tarps can be pulled back and aeration started. Fans placed near the bottom of the hold during loading can be started and will save hours of aeration time. Aeration is complete when the gas concentration in all areas is below 5 ppm. for methyl bromide or.3 ppm with phosphine.

Bulk grain treated with phosphine

The only practical fumigant to penetrate large amounts of bulk grain is phosphine. It can do an excellent job if the gas penetrates to all areas in sufficient concentrations. Penetration is obtained through long exposures or through placing the tablets throughout the grain mass with probes. Although many factors can influence the time required for phosphine gas to penetrate throughout a hold, some rough guidelines are:

- Barges - Allow at least 4 days
- Small freighters - Allow at least 7 days
- Large ships - Allow 14 days and use ducts if possible

The large amount of phosphine required (58./tonne) should never just be placed on top of the grain. Particularly if the grain was loaded during wet weather, the release of the gas into the head space can be much faster than its penetration into the grain. The standard dosage could give over 1000 ppm if the gas was evenly dispersed and this would not be dangerous. However, a fast release could cause gas readings in the head space of over 17,900 ppm which could a fire or explosion.

If the tablets are evenly probed into the grain mass, there should not be a problem. For even distribution by probing, try to probe once per square meter. Probe as deep as possible, about 2-3 meters. Apply 4-5 tablets per meter is recommended but probably not practical under ship fumigation procedures. The purpose of placing only a few tablets each time the probe is withdrawn a little is not only to disperse the tablets for better gas distribution but also to separate the tablets to avoid high concentrations of gas.

All tablets should be at least 30 cm. under the surface of the grain. After the probing, the hatch cover should be replaced and covered. The three tarps offer used can provide a reasonably good seal if they are in good condition. If there is leakage, additional sealing will be necessary. A trained person with gas measuring equipment and respiratory equipment must accompany the ship and the Captain must be prepared to take whatever precaution is necessary to protect the health of people on board. Monitoring of the gas level should include areas adjacent to the holds as gas can leak into occupied or rarely occupied area and be dangerous.

Safety practice in ship fumigation

The first step in any intransit ship fumigation is to examine the ship with the Captain or a designated ships officer to establish if the ship's holds can be safely fumigated intransit. The holds should be examined for obvious points of leakage but unusualy conditions must be considered. There may be cross connections from the bilge to drains in the galley or other areas. Gas has been known to follow this path.

Smoke detectors draw air from the hold to a sampling point at the bridge in some ships. This could bring gas from the hold to the bridge and at least one fatality may have occured from this. It may not be safe to discontinue the use of the smoke detector but additional safeguards would be indicated.

Some crews like radios in their quarters and will rig elaborate antennas. They have been known to drill through a common wall between a hold and their quarters in order to get the antenna wire to a distant point. This can permit gas leakage into the crews quarters. Doorways to the crews quarters are often left open even when they are next to a fumigated hold. Again leakage can occur.

Other areas are occasionally occupied by the crew and have been known to have high gas concentrations. Paint lockers and rope lockers may have hatches to the holds that are fumigated. Crews will sometimes sit on the hatch covers or sleep on the hatch covers and can be overcome by very small amounts of gas leakage. All holds have vents and even though these may be well sealed at the time of sailing, the seals may deteriorate later. A trained person must be on board the ship to measure the concentration and watch for unusual problems. He may be a ships officer or may be part of the fumigation crew. The key thing is sufficient training and equipment.

Fumigations for flat grain storage

Flat grain storage structures can be built quickly and at a low cost per ton of storage but they have several problems. Even with aeration ducts and the fans running regularly, it is difficult to maintain uniform moisture in the grain. Fumigation is even more difficult and most fumigators will agree that it is the most difficult type of fumigation to achieve 100% control. This is due to the tremendous amount of surface compared to the total amount of grain. The contrast would be silo fumigations where there is only a limited surface at the top compared with the total mass.

It is the large surface area where the gas is lost on most fumigations. There is also gas loss through the metal walls as the gas can leak between the corrugated sheets where they overlap. The problem is compounded by poor storage practice. The grain is rarely levelled so covering it with a tarp is difficult. There apparently has been a tendency in the past to take some short cuts from suggested procedures. The grain wasn't levelled first and the tablets were "walked in" rather than probed deeply into the grain.

The grain was probably not covered with a tarp to hold the gas within the grain mass. Some tablets were placed within the aeration ducts and the fans were run for about 15 min. Unfortunately all of these practices are wrong and could even be dangerous or could contribute to resistant insects. The natural air currents in the grain would probably be up-ward so that very little of the gas would go below the point of application. It is very easy for the gas to escape through the roof vents and at the roof wall juncture.

The fans would have very little effect on the phosphine in the first 15 minutes since it would not reach its peak of release for 24-48 hours and the release is slow then. Flat grain fumigations are done in many countries and can be reasonably successful but there is no substitute for the hard work it takes to do a good job.

The fumigation should be started by preparing the grain and the structure for the fumigation. Two crews of workers should be selected. One will be the applicators and the others will spread a plastic sheet over the grain and tuck it in at the edges and pin together successive sheets. There should be enough people on the application crew so that they can line up across the width of the grain and have no more than a meter between each worker.

The line of applicators can then move across the grain probing phosphine into the grain every meter. This will give good horizontal coverage even if the tablets are probed only one or two meters deep. The other crew will follow the applicators spreading out the plastic and joining sheets by overlapping and pinning with bamboo pegs. The two crews should be able to work fast enough to move through the area before appreciable gas has been released but gas measurements are still advisable.

After the tablets have been probed and the grain covered, gas loss can be further reduced by sealing any roof vents with plastic bags and sealing doors and windows with plastic sheets or tape at the joints. At least 7 days exposure should be used but the plastic should be removed within two weeks or sooner if the grain has a high moisture content. Sweating and mold can occur otherwise. The people removing the plastic should be protected with gas masks if the concentration is above .3 ppm. The gas level can often be reduced below this at the end of the fumigation by merely leaving the doors and vents open for a day before the plastic removal.

Training methods

- Basic training
- Groups to receive training

- Subjects to be covered with each group and why
- Suggested training techniques
- Training materials available
- How to prepare additional training material
- Subjects to Study
- Economic loss due to insects and molds
- Important Insect species
- Insect control through Integrated Pest Management
- Non-chemical controls
- Pesticides
  - a) Principles of Fumigation
  - b) Fumigants available in Thailand
  - c) Sealing techniques for fumigation
  - d) Monitoring techniques Respiratory protection
  - e) Insect resistance problems
  - f) Use of residual pesticides and grain protectants
  - g) Safety in use of pesticides
- Vertebrate pests and their control
- Mycotoxins and their control
- Warehouse practices and Sanitation
- Government regulations

Problems in fumigation

- Commodity pricing can cause fumigators to take some short cuts and avoid any costs that can be eliminated.
- The gas-proof sheets are patched many times before they are replaced. Tape used to seal cracks is narrow 2" rather than 3". Insufficient time is spent

in proper folding of tarps at the floor to get the best possible seal.

- At least on export grain fumigations, an officer of the Dept. of Agr. monitors the dosage so that the correct amount will be applied. We can only speculate on dosages at other fumigations. The accent seems to be on the dosage rather than the concentration throughout the exposure. Fumigators do not return every few hours to take gas readings.
- Leak detectors are not used even though this would not add substantially to the cost. Small leaks in the seal are probably often missed.
- Fumigators sometimes do not have access to the many training meetings that occur in many other countries. Training is often in-house and may not be sufficient.
- Most books and periodicals on fumigation are written in languages other than the local language. This makes home study impossible for many fumigators.
- Most pesticides, equipment and safety materials must be imported at a high cost. Instructions will usually be in English and require explanation by informed professionals.
- Exporters and others often specify the type of fumigant to use. They may not understand the limitations or safety problems of that choice.
- Godown managers do not realise the importance of storage practices which makes good sealing impossible.
- Ships and barges are often in poor condition making safe and effective sealing expensive time consuming and generally difficult.

- Minimum standards for various types of fumigation should be printed in local language. Plant Quarantine should refuse to issue certificates for fumigations deemed by their officers to be possibly dangerous or ineffective.
- Training sessions should be held in the local language.
- Godown personnel should be trained to store properly
- Exporters should be informed about the limitations of each fumigant.
- Equipment should be developed that would permit recirculation of methyl bromide in barges or similar areas.
- Leak detectors should be used on all fumigations and enforcement is suggested wherever a government officer is required to witness a fumigation.
- Establish an import requirement that all pesticides, equipment and safety devices must be accompanied by adequate instructions in the local language.
- Give government officers the right and duty to disapprove vessels that in their opinion might create a safety hazard for in-transit fumigation.
- Stress to exporters and other purchasers of fumigation the high cost of infestation at the port of arrival compared with a modest increase in fumigation cost to do a better job according to new standards.
- Encourage fumigators to sell additional services with the fumigations. Residual spraying, space treatments, and even inspection services can be cost effective.

- Set up a certification programme for fumigation supervisors with written exams and recertification by exam or attendance at training sessions.
- Restrict sale of dangerous fumigants to persons that have passed the writter exam.
- Sponsor attendance by key Dept of Agr. Officers at fumigation training meetings in other countries.

Field test of fumigations

Determine efficacy of present barge fumigations with monitoring lines checked every 4-6 hours on a Fumiscope or other device. Test insects will help evaluate control although there will probably be a good natural infestation. This test could then be repeated by making minor changes such as using pallets on the bottom and a higher expansion chamber on with multiple release points for the gas. We can only speculate about present hazards. Gas testing for comparison with suggested Threshold Limit Values should be done on the following:

- Intransit barge fumigations
- Intransit ship fumigations
- Godown tarp fumigations with nearby workers
- Fumigated containers that may be stored below deck without aeration.
- Tobacco sample room fumigations next to offices.

A demonstration godown should be selected and given the entire proposed programme including fumigation of all stacks at one time and application of residuals followed by a space treatment. Godown personnel would also be taught how to inspect and monitor the control. This would then be visited by other interested parties to see that control is possible. Try to obtain the total weight of grain bought per year by one godown and compare with the total weight sold. An attempt then should be made to determine the sources of the loss and evaluate the significance to a good pest control programme. The efficacy of various tarp treatments should be made by sampling before and after such typical fumigations as:

- Fumigation tarp bunched at the bottom permitting leaks
- Fumigation tarp held down with bags instead of sand snakes
- Fumigation tarp not sealed to floor near walls or posts
- Fumigation tarps properly sealed to smooth out tarp held with overlapping sand snakes.

Recirculation of methyl bromide

Export grain must arrive insect free at the port of destination to receive a good price. Fumigations inside godowns are subject to reinfestation. Ship fumigations in transit can be hazardous and may not be effective due to the depth of the holds. Barges are often used to transport grain from the godowns to the ship. They present a sealable area isolated from probable reinfestation but there have been problems in obtaining insect control during barge fumigations. The 24 hour holding period of the barge at the dock is not long enough for effective fumigation with phosphine and there is strong reluctance to pay for longer demmurage.

The tightly packed bags placed directly on the floor of the hold, impede fumigant circulation. There is no use of fans and only limited expansion chamber in the top. Although we saw multiple release points used in ships holds, we never saw more than one release point used on barges. This type of fumigation can be expected to be less than perfect. Recognising the success of recirculation of Methyl bromide in installations such as the Controlled Circulation Chamber for Orchids, we propose to test a modification of the barge fumigation procedure.

Instead of placing the first layer of bags on the bottom of the hold, rows of pallets should be placed with all 2 x 4 runners lined up with the ship so that air channels would run fore to aft. The rope locker would be used as one plenum chamber. The door from the hold to the rope locker would be left open. A board would be sealed to the hatch opening and a hole in the board would be connected to a gas powered air pump. The air would then be ducted to the top of the hold and the tarp would be sealed around the duct as it is around the entire perimeter. The gas would be injected into the air stream where it would mix with air in the plenum created by the gas proof sheet held off the cargo by pipe hoops.

The gas/air mixture would settle through the bags both by gravity and by the pull of the blower drawing air from the hold into the rope locker. The fan should run for the entire 24 hours and then again during aeration after the tarp had been removed. This will provide better safety for workers that must enter the hold to unload the bags by hand. The pipe hoops may need to be lengthened to form, a larger expansion than normally exists and to allow for the loss of space taken up the pal lets.

Test lines should be placed at the top, middle, and bottom at the front, middle and rear as well as the rope locker and crews are to evaluate the concentration. The fumiscope should be used for the readings. The crew should not be allowed on board during the fumigation but could be on the dock if air sampling is done. There is always a hazard to the crew if they stay on a barge that is being fumigated but even a small leak will be serious during recirculation. The size of the blower must be kept small enough to not put too much strain on the seal of the tarp put should be enough to assure proper circulation of the air gas mixture.

Detection of Gaseous Concentrations

Detection of gaseous concentrations is an essential prerequisite for both operational and safety reasons when any fumigation is attempted. The requirements are the determination of excessive leaks from the fumigation enclosure (sheeted stack) or building, the concentrations achieved during the course of the fumigation, the concentration reduction during aeration, and protection of the fumigator from potential hazards associated with the fumigant. The threshold limit values (TLV's) have been revised with respect to the common grain fumigants in present use, which emphasises the need for more accurate detection of environmental contaminants for operator safety.

Detection equipment must be used in preference to odour, even though some fumigants can be detected (not accurately;) this way. Phosphine possesses a garlic or carbide like odour (due to probably diphosphine, since pure phosphine is odourless) but may not be apparent due to preferential sorption on grain. Methyl bromide is odourless at low concentrations, and most formulations contain chloropicrin as a warning agent. Again, atmospheres containing methyl bromide may not be apparent to the fumigator due to different sorptive characteristics of chloropicrin (BP 112°C) as compared to methyl bromide (BP 3.6°C).

Detector Tubes

These are sealed glass tubes which, after the tips are

removed, are inserted into a device and air is drawn through the tube. The fumigant then reacts chemically with the reagent coated granular filler and the colour change is measured directly on the tube or with a calibration chart. There are various types of pumps, such as Drager, Auer, Kitagawa and MSA. Detector tubes are available for phosphine, methyl bromide, carbon tetrachloride, carbon disulfide, hydrocyanic peid, (acrylonitrile, ethylene oxide, carbon dioxide, ethylene dibromide, ethylene dichloride, oxygen, and other environmental contaminants.

Drager multi gas detector

The pump consists of a simply designed, extremely strong bellows which requires practically no maintenance. The pump is easily operated by one hand and sucks in exactly 100 cm^3 (100 ml) of gas per stroke, and measures the gas sample at the same time. The same bellows pump is used for the dust sampler to suck in the air sample for the analysis of dust concentrations. The Drager tube is a glass tube fused at both ends and is chemically stable in this condition with a normal shelf life of approximately 2 years.

Both ends of the Drager tube are opened in the ampoule breaker eyelet by a "rotating, breaking motion", the eyelet being located next to the loop which retains the arrestor chain. The break off husk can also be used to open the ampoule. It prevents glass splinters from falling, and is provided in the carrying bag of the Gas Detector. The husk should be emptied from time to time. The open tube is then inserted into the pump head so that the arrow of the tube points towards the pump and the writing surface towards the air inlet. The tube must be located firmly and tightly in the pump head so that no bypass air can be drawn into the pump.

During the test, the pump body is gripped by the working hand so that its base plate, which carries the pump head and takes the form of a recessed handle, is firmly located against the palm and the ball of the hand. The pump is held between the thumb and the base of the forefingers. The four fingers rest loosely on the front plate. The pump is operated by compressing the rubber bellows as far as the stop and releasing it. When the pump is compressed, the air escapes form the bellows through the eyelet valve in the front panel and not through the tube, because this has a much higher flow resistance than the outlet valve.

The suction of the pump begins with the opening of the four fingers which previously rested on the front panel. The fingers should be completely released from the panel, but the pump will still be firmly located between the thumb and the base of the forefingers. The compression springs within the bellows which are tensioned when the bellows are compressed, expand and the outlet valve is closed by the vacuum created in the bellows. The air then flows into the bellows, through the tube while the bellows expands to its original volume. The air volume drawn through the tube is determined by the dimensions and the stroke of the bellows.

The end of the suction movement is reached when the arrestor chain is fully tensioned. The air velocity in the tube, which is decisive for the accuracy of measurements, is thus determined entirely by the spring force and the resistance of the Drager tube charge, which, is fixed during manufacture. The number of strokes prescribed for the various Drager tubes must be observed at each test (Table 2).

Auer Gas-Detector(R), Auergesellschaft, GMBH, FRG

The Auer Gas-Detector(R) consists basically of a suction

Table 2

Gas and vapours	Drager tube used	Order code (1 pack)	Measurement range (20°c; 1013mbar):	No.of strodes	TLv
Perchloroethylene (5)	Perchloroe-thylene 51a	6.7 26699	5-50 ppm	10	1000 ppm
(Tetrachloroethylene) (5)	" 10/b CH 30701		10-500 ppm	3	100 ppm
	" 0.1%1a 67 28021		0.1-1.4% vol.	5	-
1,1,1-Trichloroethane (methyl chloroform)	Trichloroethane 50/C	CH 21101	50-600 ppm	10	350 ppm
Phosphine	Phosphine 0.11a	CH 31101 0	.1-4 ppm	10	0.3 ppm
	Phosphine 501a	CH 21201	50-1000 ppm	3	-
			15-300 ppm	10	-
			150-3000 ppm	1	-
Phosphoric acid esters	Phosphoric acid esters 0.021a	67 2846 0.02	ppm DDVP	5 DDVP	0.1 ppm

ball bulb and a pump body into which the detector tube is inserted. When the suction ball is compressed, a negative pressure is produced which draws the air to be sampled through the detector tube, the chemical composition of the indicating material being specifically developed for the gases and vapours being monitored showing a characteristic discolouration on their presence. In most cases, a specific sample volume is drawn through the detector tube and the concentration of the substance to be detected is read from a scale according to the length of the stain or is evaluated by means of a table.

Some detector tubes have only a very short discolouration (annular indication). In these cases, the sample volume needed to obtain the discolouration is equivalent to the concentration. Detector tubes have been developed for specific gases and vapours, or for broad spectrum application for a number of gases.

The Auer Gas-Detector is simple with only 10 parts (locking screw; counting device, disc; washer; spring pin; spring; pump body; bushing; tube breaker; valve washer; and aspirator bulb for model 69, and can be taken apart and reassembled very quickly. The use of plastic and rubber makes the device largely resistant to corrosive gases and vapours; and combined with sturdy construction, virtually no maintenance is required. Where AUER Toximeter (TX), an electronically operated automatic pump for detector tubes is recommended.

Instructions for operation

Break off both tips of the detector tube in the tube breaker on the pump body of the Auer Gas Tester with a rotating movement. Insert detector tube tightly into the pump body in the direction of the arrow printed on the tube (unless otherwise specified in the special instructions). Set disc counting device to the number given in the instructions for use of the detector tube used. One hand operated, hold neck of suction ball between the tips of the index and middle fingers, apply thumb broadly to the operating ring at the bottom of the suction ball.

Compress suction ball firmly in the direction of the pump body up to the stop. Do not try to squeeze out the residual air contained in the sides of the suction ball. Take off thumb and allow suction ball to expand keeping the detector tube for another 5 seconds in the sampling area (unless otherwise specified). Repeat sampling procedure as often as indicated in the instructions for use of the detector tube.

Move the disc counting device one by one digit before every stroke. The disc counting device is designed to engage at 0, which shows the user that the test is completed. Some detector tubes give off acid vapours during testing. Although the device is extremely resistant to these vapours, it is advisable to flush it after each use with a few blind strokes. Evaluate detector tube reading as described in the instructions for use.

Phosphine detector tubes

PH3-0.1: Used for detecting phosphine in air or technical gases over a range from 0.11Q0 ppm. For concentrations greater than 50 ppm and for checking TLV.

PH3-50: Used for detecting phosphine in a range from 50-2000 ppm.

The detector tubes have two scales, or calibrated for use with AUER Gas-Tester hand pump (G.T.), the other calibrated for use with the AUER Toximeter (TX).

Reaction of phosphine with a silver compound and the formation of metallic silver. The indicator layer (white) is progressively discoloured zone for zone to brown in the aresence of phosphine. The concentration indicated only applies for the selected location, and only for the moment in time the measurement was taken. Changes in concentration (evolution, gas leakage, formation of vapour clouds, etc) must be taken into consideration.

Accuracy:

Length of stain, more than 10 mm:t 10 to ± 15% standard deviation

Length of stain, less than 10 mm:+ 15 to ± 25% standard deviation

Measurement

Tube type : Pump type	PH3-0.1		PH3-50	
	:Gas tester (GT):	Toximeter (TX):	Gas Tester (GT):	Toximeter (TX)
Connecting detector tube	Arrow points to-wards pump	Arrow pointing away from pump	Arrow in detector tube indicates flow direction (towards pump)	Insert detector tube into rubber nozzle of pump. Arrow indicates flow direction (away from pump)
Range of measurement	10 strokes :0.1 to 10 ppm	Switch position 10: 0.1-10	1 stroke (1 H) 50 to 2000 ppm	Switch position 1: 50 to 2000
	Take reading on 10 stroke	ppm Switch	For evaluation use GT scale	ppm For

	scale (GT) 1 stroke: 1 to 100 ppm Take reading on 10 stroke scale and multiply by 10. (GT)	position 1: 1 to 100 ppm		evaluation use TX scale
Suction time	20 seconds per stroke including waiting time	Switch position 1: approx. 25 seconds Switch position 10: approx. 2.50 seconds	Approx. 25 sees. per stroke	Switch position 1: approx 25 seconds

	Gas Detection Equipment
Phosphine	Detector Tubes
Methyl bromide	Detector Tubes
	Halide Gas Detector
	Thermal Conductivity Analyser
	Interference Refractometer
Carbon tetrachloride	Detector Tubes
	Halide Detector
	Thermal Conductivity Analyser
	Interference Refractometer
Carbon disulfide	Detector Tubes
	Thermal Conductivity Analyser
Hydrocyanic acid	Detector Tubes
	Colour Indicator
Sulfuryl fluoride	Thermal Conductivity Analyser

9

Insect Control

Integrated Pest Management (IPM)

Pest control measures have to be integrated into an operational system, be it large or small in scale, if they are to be effectively applied. This is a basic principle, not a novel concept, but it connects well with the modern idea of Integrated Pest Management (IPM). The use, in that term, of the word 'management' is appropriate, especially with regard to the need for the integration of pest control measures into management systems, but it should be remembered that the two words, 'management' and 'control', are almost synonymous. The fundamentally important emphasis should be placed upon the word 'integrated'.

Integrated pest control can be defined as the acceptable use of practicable measures to minimise, cost-effectively. the losses caused by pests in a particular management system. For the measures to be cost-effective they must be appropriate to and acceptable into that system. They may be simple or complex but they must suit the system objectives and its technical capabilities. Furthermore cost-effectiveness requires that all costs and benefits, including sociological and environmental effects, should have been taken into account. The term integrated pest management is used to imply that a flexible and technically informed approach is also required.

In defining this term it may be considered necessary to specify the inclusion of scientific and cost-effective pest monitoring procedures which permit judicious adjustments to the timing, choice and intensity of control actions. It may also be advisable to point out that specific pest control measures, as distinct from general crop or commodity husbandry practices, should generally be omitted unless the circumstances warrant and permit their cost-effective inclusion. Insect pest management for stored grains, like preharvest pest management, can thus be seen, historically, as a traditional approach in which good husbandry is the primary requirement.

Unfortunately, it must also be acknowledged that the advent of readily available, relatively inexpensive synthetic insecticides has led to considerable over-dependence upon these hazardous tools with, in some cases, a consequent neglect of basic good husbandry. The current emphasis upon integrated pest management is, in effect, a reassertion of the need to put traditional good husbandry in place as the fundamental basis of pest control. In grain storage, as with other durable agricultural products, it is good commodity management and good store management which are the major prerequisites (Tables 1 and 2).

The various options for more intensive insect pest control, which are also listed in Tables 1 and 2, include several which are themselves based upon traditional concepts of pest management. Thermal disinfestation, cooling and hermetic storage are examples. These latter two methods are also examples of the opportunities, provided by the process of storage, to manage the generally enclosed storage environment in such a way that insect pests are prevented from multiplying or, as in efficient hermetic storage, effectively eliminated. Preharvest problems of insect pest control are rarely, if ever, so easily managed! Control of the storage environment is thus an essential element in grain storage pest management.

It involves, primarily, the controls on in-store climate and infestation-pressure which can be achieved by technically sound store design and construction. Equally important, however, is the climatic control attainable by scientific management of the commodity to ensure that the stored grain is itself both dry and cool when loaded or, in ventilated stores and bins with aeration equipment, that the storage procedure achieves drying and cooling sufficiently rapidly. In a fully loaded store it is the stored grain itself which largely determines and stabilises the temperature and humidity conditions in the store. Commodity management can also control, to a considerable extent, the initial insect infestation level in the stored grain.

However, in tropical countries, where preharvest infestation by storage insects is hardly ever completely preventable, the ideal of loading insect-free grain into the store is not often attainable. Special facilities to completely disinfest the grain before loading may not prove cost-effective. The common alternatives, if early disinfestation is required, are to treat the grain, at intake, with a suitable admixed insecticide or to disinfest the loaded grain by in-store fumigation. Control of grain quality before storage, to minimise the intake of heavily infested and badly damaged or uncleaned grain, is feasible and is commonly practiced to a considerable extent.

Even at the small farm level it is possible to segregate the crop at harvest, especially with maize on the cob and unthreshed sorghum and millet, selecting relatively undamaged material with good storage potential and setting aside the more evidently infested or otherwise damaged material which, if there is no other option, can at least be used first. By such means, the rate of deterioration due to insect infestation can be considerably retarded in the main stock of stored grain. There is little doubt that some subsistence farmers use this form of commodity management fairly effectively.

Certainly, one can sometimes observe onfarm grain stocks, that have received no special insecticidal treatment, with relatively little insect damage after several months storage at an ambient temperature that would permit the rapid increase of any well-established initial insect population. Scientific approaches to grain storage pest management, having regard to grain storage as a part of the food production and distribution management system, have sometimes referred to the biological ecosystem concept as a means of comprehending grain storage processes and problems.

In a recent contribution Dunkel has applied ecosystem principles in a broad analysis directed towards an improved understanding of physical and biological interactions including socio-economic factors. The purpose was to generate improved understanding of the stored grain ecosystem and to identify objectives for future postharvest research. This treatment of the subject should serve to enhance the growing awareness of storage as a system within a system and to stimulate systematic and objective analysis of grain storage problems. Whether or not one prefers to use the term 'ecosystem' is less important and this somewhat academic term should not be allowed to obscure the main issues.

Table 1. Prerequisites and options for on-farm storage pest management

Essential Basic IPM	Optional Additional measures
Site and store managemen: (protection from birds, rodents and weather plus basic hygiene)	Maintenance of conditions favourable to natural control: by cooling (where feasible) by insect parasites, pathogens, etc. and/or Thermal disinfe:tation by solar heat and/or Treatment with traditional additives (if sufficiently available and effective) or
Commodity management	Treatment with synthetic insecticides (if suitable formulations sufficiently available and effective)orHermetic storage (pits or metal drums, etc.)

Table 2. Prerequisites and options for storage pest management at main depots

Essential	Optional	
Basic IPM	Disinfestation	Prevention of reinfestation
Site and store management (protection from birds, rodents and weather plus basic hygiene)	Insecticide admixture Fumigation! or Thermal	Provided by the treatment Residual insecticide sprays? Physical protection! (Sheeted stacks or packaging) or
Commodity management	Irradiation	
with bulk storage if	Hermetic	
appropriate	Controlled atmosphere space	Insecticidal treatments! Provided by the system Provided by the

	system
	Provided by the system

The integration of various control techniques, within the framework of integrated pest management, has become a focus for research in stored products work. The importance of a multidisciplinary approach to stored grain research has also been stressed. This is very valid but it is useful to recall that a great deal of the research done in the past has been of this nature. It was pointed out, at a stored products pest management symposium held in 1978, that the need for an interdisciplinary approach is generally well-known.

Entomology, mycology, chemistry, engineering and food science are commonly involved, but effective integration of technical solutions is often lacking; possibly because some of the more pragmatic disciplines, notably economics, sociology and business management, are not always sufficiently involved. It is the interface between the research team and the storage managers, whether these be individual farmers or a large storage organisation, which may sometimes be the most crucial barrier to progress. A clear perception of the need for solutions which can be integrated into the management system, because they meet the business objectives and can be accommodated within existing management capabilities, is probably the most important requirement.

Modern theories of pest management have also generated the concept of economic control thresholds (ECTs). An ECT is most simply defined as the level of pest damage which justifies, in cost/benefit terms, the expenditure of resources upon control actions. It is always a variable threshold because the costs and benefits of any action will depend upon the situation and its circumstances. An ECT is situationspecific. This is especially true, and not only for postharvest pest control, when one considers the great differences in opportunities and constraints between the small farm level in developing countries and more sophisticated levels of operation. Nevertheless, it is possible to generalise to some extent. For insect control in grain storage the ECT is likely to be at or very close to zero:

a. where consumer demands place a high value on freedom from insect damage and/or freedom from any sign of insect infestation;

b. where there is a definite intention to store grain for a protracted period, in which substantial insect damage can be predicted, and where the eventual market is not insensitive to loss of quality.

In both cases the assumptions are made that at least one potentially cost-effective package of control measures is available and that the package provides for sustained control: including efficient prevention or control of reinfestation. In most sophisticated situations, where consumer quality standards are likely to be high, these will be valid assumptions. Otherwise the ECT could not be zero and the required quality standards would have to be met by reconditioning the product.

The losses, including reconditioning costs, would then have to be borne by the system or the consumer. A real economic loss would have occurred, masked by marketing tactics. Such events do, undoubtedly, take place. Where grain is being stored for domestic use, or for eventual sale in an uncritical market, the ECT may be well above zero. This can be true even when all parameters of economic damage are considered because, in fact, the true economic significance of a low percentage of insect damaged grain may be virtually nil.

The actual loss in real food value is likely to be negligible and the market value of the associated weight loss, if saved, may not offset the full costs of pest control actions. The need to minimise the cost of insect pest control is a major factor militating against the extensive use of some of the more 'environmentally friendly' measures which might otherwise be preferred to the continued use of suitable fumigants and contact insecticides. However, it must also be acknowledged that chemical pest control measures, in grain storage, are often not only the cheapest but also the most reliably efficacious of the possible options. Where this remains true, and for as long as such treatments are accepted as generally safe, one important requirement for IPM in grain storage will be to ensure that chemical measures are recommended only where they can be safely and efficiently used and only when they are economically justifiable.

The economic justification of control measures, including where necessary the use of chemical pesticides, should take account of all costs and benefits. Many of these may be assessable only in subjective terms that are greatly dependent upon local attitudes and sensibilities. However, measurable losses of quantity and certain quality parameters can be objectively determined. A manual of methods for the evaluation of postharvest losses is available and provides useful information, subject to a need for modification of some methods in particular circumstances. A critical review of the methodology gives further guidance based upon experience gained, since 1978, from field work in many developing countries.

Choices among methods, several of which have become subjects of considerable controversy, should always be made with due regard to the actual circumstances and the prevailing objective. There is no single 'best method' for all circumstances and, for practical purposes, operational facility and repeatability are generally more important that fine precision. From the storekeeper's viewpoint a demonstrable loss reduction from, say, 5% to 4%, however statistically significant, may be of little or no practical importance; whereas a reduction from 10% to 5% (or from 2% to 1% in a sophisticated storage system) may be of considerable interest: provided, always, that the demonstrated reduction can be achieved in routine practice.

Insect Control in Hot Climates

The insect species of most importance in the tropics as pests of stored foodgrains, including pulses (legume grains), are listed in Table 3. Recognition of these common major pests, at least to the genus level, is not difficult for a trained inspector and there are many useful recognition charts and some excellent keys to specific identification.

Table 3. Important insect pests of tropical stored grains or grain products

COLEOPTERA:	ANOBIIDAE	Lasioderma serricorne (F)
	BOSTRICHIDAE	Rhyzopertha dominica (F)
	BRUCHIDAE	Prostephanus truncatus (Horn).
	CUCUJIDAE	Acanthoscelides obtectus (Say)
	CURCULIONIDAE	Callosobruchus spp.
	DERMESTIDAE	Zabrotes subfasciatus Boheman
	SILVANIDAE	Cryptolestes spp.
	TENEBRIONIDAE	Sitophilus oryzae (L)
		S. zeamais Motschulsky
		Trogoderma granarium Everts

		Dermestes spp.
		Oryzuephilus surinamensis (L)
		Tribolium castaneum (Herbs)
LEPIDOPTERA:		Sitotroga cerealella (Olivier)
	GELECHIIDAE	Ephestia cautella (Walker)
	PYRALIDAE	Plodia interpunctella (Hubner)
		Corcyra cephalonica (Stainton)

The grain weevils (Curculionidae) are well-known as major primary pests of stored cereal grains. They are able to establish themselves on whole, undamaged grains of maize, sorghum, rice and wheat so long as the grains are not exceptionally dry. However, Sitophilus zeamais is the dominant species on maize while Sitophilus oryzae is dominant on wheat. Neither species is a significant pest of millets and other grains that are too small to permit the full development, within a single grain, of the weevil larva.

The bostrichid beetle Prostephanus truncatus (the Larger Grain Borer) is a highly destructive primary pest of maize, especially maize stored on the cob. This insect is now established in several East and West African countries following recent accidental introductions from its previously more limited indigenous range in meso-America. The Lesser Grain Borer (Rhyzopertha dominica) is more cosmopolitan and is well-known as a destructive pest of most stored cereal grains including millet. It is not generally common on maize. The bruchid beetles listed in Table 3 are the only significant pests of stored pulses.

Acanthoscelides obtectus and the less cosmopolitan Zabrotes subfasciatus are generally restricted to dry beans (Phaseolus vulgaris) while Callosobruchus spp are generally restricted to the other legume grains, notably cowpeas and mungbeans (Vigna spp). The anobiid Lasioderma serricorne is sometimes reported as a significant pest of stored beans and other pulses but it is not a primary pest of foodgrains. It is, however, a considerable pest of cereal-based animal feeds, wholemeal flour and high-protein milling offals. Apart from the one exception noted above (L. serricorne) all of these beetles are primary pests in the sense that they can initiate major damage to the grains and most of them are also able to commence their attack in the field before harvest.

The other beetles listed in Table 3 are generally regarded as secondary pests which can be of major importance on grains previously damaged either mechanically or by other insects. The dermestid Trogoderma granarium is exceptional in that it can cause major primary damage but rarely occurs as a primary pest except in arid climates, or on very dry grain, where other primary pests are inhibited by the dryness. Of the moths (Lepidoptera) listed in Table 3 only the gelechiid Sitotroga cerealella is able to cause substantial primary damage to the grain kernel.

Like the grain weevils it can also infest the grains before harvest and like the Lesser Grain Borer it is a considerable pest of millets as well as all the larger cereal grains. The other moths listed are warehouse moths, and the rice moth Corcyra cephalonica is included here although it is less commonly abundant on other cereal grains. The larvae of all three species can do substantial damage as secondary pests. They can also attack the whole grain at the site of the embryo, which is typically excised completely, and may thus be of special importance as pests in seed grain stores.

Many other insects may occur quite commonly and sometimes abundantly on stored cereal grains especially when they are underdried or have been heavily infested by the major primary and secondary pests. Some of these are listed in Table 4. This list, although quite extensive, is not exhaustive. Most of the species included are capable of doing some damage to the grain and while several are unable to thrive in the absence of mould growth, or are most commonly predators and scavengers rather than grain feeders, their presence would be unacceptable to many consumers. Evident signs of their previous activity, in the form of insect fragments and waste matter, will be equally unacceptable where high quality commands a premium price.

Table 4. Insect species found on underdried stored grain or grain residues

COLEOPTERA:	ANTHRIBIDAE	Araecerus fasciculatus Degeer
	BOSTRICHIDAE	Dinoderus spp.
	BRUCHIDAE	Bruchidius spp., Specularius spp.
	CLERIDAE	Necrobia rufipes Degeer
	CRYPTOPHAGIDAE	Thaneroclerus buqueti Lefevre
	DERMESTIDAE	Henoticus californicus (Mann)
	LATHRIDIIDAE	Cryptophagus spp
	MYCETOPHAGIDAE	Attagenus spp., Dermestes spp.
	NITIDULIDAE	Corticaria spp., Lathridius spp.
	OSTOMIDAE	Typhaea stercorea (L)
	PTINIDAE	Carpophilus spp.
	SILVANIDAE	Tenebroides mauritanicus (L)
	TENEBRIONIDAE	Ptinus spp.*, Trigonogenius spp.,

		Gibbium spp.
		Cathartus quadricollis (Guerin)
		Alphitobius spp.,
		Gnatocerus spp.
		Palorus spp.
LEPIDOPIERA:	OECOPHORIDAE	Endrosis sarcitrella (L)
PSOCOPTERA:	LIPOSCELIDAE	Liposcelis spp.

The psocopteran (psocid) species listed in Table 4, which include the familiar 'dust lice', sometimes mistaken for mites (Acarina), have received more attention in recent years than previously; both in the tropics and in temperate countries. In the latter, they have attracted attention as occasional pests of various commodities including skimmed milk powder. In the humid tropics they sometimes occur as troublesome pests of cereal grains, notably milled rice, on which they are sometimes very abundant. It has been shown that a psocid population can feed and multiply on damaged or imperfect cereal grains. When milled rice is exposed to very dense populations the feeding damage, on some individual kernels, can be massive.

Whether or not such damage occurs sufficiently extensively to constitute an economically significant weight loss in storage practice has yet to be demonstrated. However, it has been shown that measurable weight losses can occur, especially on under-milled rice, and that the protein and lipid content of the infested rice may be reduced. The nuisance-value of these small insects, when they occur as dense, swarming populations in warehouses, is also considerable. Furthermore, surface scarification due to psocid feeding may improve the apparent whiteness of infested milled rice and thus obscure the visible evidence of mould growth and possible contamination by mycotoxins.

Status of Pest

The status of any particular insect pest may vary between different commodities, different varieties of the same commodity, different climatic regions and agro-industrial systems and between different socio-economic groups. It is affected by the form in which the commodity is stored, by the environmental conditions and by consumer attitudes. As with the psocid problem, it may also be affected by the sensitivities of store supervisors and their work-force. Pest status may also vary between biotypes of the same insect species due to differences in the capacity to cause grain damage or to adaptations to other foodstuffs.

As an extreme example of this, although grain weevils are usually insignificant as pests of stored pulses there are biotypes of S. oryzae that multiply successfully on stored split peas and on mung beans (C P Haines, personal communication: including field records). Certain strains of S. oryzAe have been noted as having greater flight proclivity than others. This species, unlike the maize weevil S. zeamais, does not usually fly very readily, although it has wings. The influence of maize varietal characteristics upon the preharvest infestation of maize cobs by S. zeamais has been much investigated.

It is clear that the cob sheath, in those cultivars which produce sheathing leaves completely enclosing the entire cob, provides considerable protection against the weevil. Storage of cobs in the sheath, which does not significantly impair the grain drying rate in ventilated cribs, therefore reduces the status of the grain weevil as a pest and will be beneficial where weevils are the main threat. Even without the sheath, grains on the cob are considerably less susceptible to weevil attack than the shelled grains.

Unfortunately, maize on the cob, especially without the sheath, is more heavily attacked by the grain moth S. cerealella. The Larger Grain Borer Prostephanus truncatus is also favoured by storage on the cob, with or without the sheath. The impact of this pest in those African countries where it has recently established itself has been dramatic for this reason as well as on account of the more rapid and destructive grain damage caused by the adult borers. The status of this new pest in maize-based farming systems in Africa, traditionally dependent upon cob storage as a modest pest management stratagem, has been such that pragmatic thresholds for control action are commonly exceeded.

In consequence, demonstrably cost-effective treatments of shelled maize grain with admixed powder formulations of suitable synthetic insecticides have been readily adopted in several regions where they were, formerly, less often used. As a further consequence, it may be that the overall insect control level has been considerably enhanced, since the recommended formulations are generally 'cocktails' of two active ingredients that can give effective control of a broad spectrum of storage insect pests. The storage of sorghum and rough rice (paddy) in the panicle, millet on the head and cowpeas in the pod also serves as a modest form of insect pest management.

However, the storage of unthreshed grains, although it may retard the build-up of infestation by some pests, does not prevent it entirely and different insects are affected in different ways. Thus, on sorghum, infestation by grain weevils is usually reduced but the grain moth S. cerealella is likely to be more successful. On unthreshed rice and millet, for similar reasons, the grain moth will have increased pest status. The low status of grain weevils as pests of millets and other small grains, previously mentioned, is due primarily to the limited grain size since the complete larval development, in these species, has to take place within one kernel.

The lesser grain borer (R. dominica) and the grain moth (S. cerealella) are not so handicapped because their larvae are able to migrate, if necessary, from one kernel to another. However, the grain moth is a low status pest on large grain bulks and tightly-built large bag-stacks, where the beetles are serious pests, because the moth is unable to move freely amongst close-packed grains to lay its eggs and the first instar larvae are unable to travel more than a few centimetres in search of food. Most storage insects, especially the important pests, are able to survive and multiply rapidly on well-dried grain.

However, grain dried to below 12% mc inhibits the development of most species to some extent and on exceptionally dry grain (<8% mc) the grain weevils, for example, are insignificant pests. The grain borers remain of considerable importance at these low moisture levels and the "khapra" beetle (T. granarium) becomes increasingly important. This insect assumes major pest status and dominance over almost all other storage insects at the very low moisture contents (down to about 4%) which equate to the extremely low humidities (< 20% rh) that characterise the most arid climates and, also, the insect ecosystem created by stored malting barley which is usually dried to this very low level and is not uncommonly imported into the tropics.

Factors Effects on Pest Management

Grain moisture content considerably affects pest status but it is not a factor which can be cost-effectively manipulated, in most situations, to achieve sufficient control of insect pests. Cost-effective drying, in common practice, can achieve control of moulds and will lessen the problems of insect infestation; in particular it will greatly reduce the spectrum of pest species. However, it will not prevent significant damage by one or more of the major insect pests. Insect development and population growth rates are more dramatically affected by temperature and here the developmental limits are more clearly defined and generally applicable.

Upper limits for development and survival vary to some extent between species, with the grain borers again more resistant than the grain weevils, but temperatures above 45°C are eventually fatal to all storage insects. At 50°C most species will die quite quickly, within a matter of hours, and complete disinfestation of wheat grain can be achieved rapidly, economically and without damage to the grain by very short exposures to air heated to 60°C. Rapid insect development occurs within a fairly narrow range of 5-10 degrees around the optimal temperature which, for most storage insects, is in the region of 30°C.

At temperatures nearer to 20°C development proceeds more slowly and population growth may be considerably reduced. At 17°C or less it is relatively negligible and pest status is consequently greatly reduced. However, even at 15°C some species are able to continue feeding, to some extent, so that grain damage may very slowly increase. Insect populations will certainly not be eliminated at these temperatures and, while grain may often be held safely in cool storage, any eventual transfer to warmer conditions will bring about a resurgence of the suppressed infestation.

Even in cold storage (at 6-9 °C) some, at least, of the important insect pests of stored grain can survive longer than one year. Insects require oxygen for respiration. Living grains, when sufficiently dry (12-13% me), are dormant and respire very little. Grain properties, including viability, are virtually unaffected in cool conditions by protracted hermetic storage. Insects, however, will use up the oxygen and eventually die. The traditional concept of sealed (hermetic) storage as a means of controlling insect infestation depends upon this.

Most storage insects will die when the oxygen in the storage atmosphere is reduced, by the insects' respiration, to 2%. With light infestation the process may take 6-8 weeks but, if airtight conditions are maintained, the infestation will be controlled and probably eliminated before serious damage is done. There is new evidence that insects may adapt to low oxygen tensions and evolve strains with considerable resistance to sub-optimal levels even down to about 1%. However, it seems very improbable that any storage insect would multiply rapidly in such conditions. Physical disturbance of grain, by turning it from one elevator bin to another, can reduce live grain weevil infestation to a considerable extent and thus retard its further development.

A more complete kill of all insect life stages can be achieved, by mechanical high-speed impact, in the entoleters included in the processing line of many grain mills. In small-scale storage it may be less easy to achieve the degree of disturbance necessary for effective control of grain weevil and other cereal grain insect pests, although very small quantities of grain held in small pots or gourds could be shaken sufficiently violently for this purpose. Bruchid pests of stored pulses may be particularly susceptible to control by physical disturbance.

In the bean bruchid A. obtectus this may be because the hatched first instar larvae require an approximately 24 hr period to penetrate the bean testa. These workers have shown that twice-daily 'tumbling' of small lots of stored beans, in partly-filled cylindrical containers, reduced bean bruchid populations by 97%. Daily sieving on a fair-sized coarse mesh can achieve similar results. However, the practicability of these proposed techniques, for routine use by farmers or traders in developing countries, is unproven.

Insect behaviour patterns may affect pest status and pest control. Examples have been given of the ways in which adult oviposition behaviour and larval feeding behaviour can affect pest status in the grain moth and pest control in bruchids. The development of infestations (pest population development) may also be affected by, for example, the diapause habit which characterises several storage insects including, most notably, the khapra beetle T. granarium. Diapause may postpone population development, usually in unfavourable conditions, and it may also impair the effectiveness of control measures; including fumigation and the use of contact insecticides as surface sprays for 'clean-up' treatments in empty storages.

Locomotory avoidance behaviour, especially in the flour beetle T. castaneum, is also of considerable interest. Storage management, which may be described as the science of cost-effective storage organisation, greatly influences pest development and control. It encompasses decisions upon the location of stores, storage periods and the quality control objectives for stored commodities. All of these have substantial implications for pest management and are components of the complex interactive network of factors affecting loss reductions in grain storage.

Modern techniques, especially those involving the application of synthetic insecticides to stored grain, are especially prone to consumer sensitivity. However, many traditional techniques face comparable problems. The use of wood ash and other supposedly non-toxic 'natural' grain protectants may not be acceptable in all circumstances, nor even to all those people at the small farm level in developing countries who are often supposed to prefer such treatments.

Pest Management Problem

The development by pests of acquired resistance or increased tolerance to pesticides is now a well-known pest management problem. It is coming to be recognised as not so much a remarkable phenomenon as an almost inevitable natural consequence of pesticide use. There is little doubt that much of the existing problem, which in recent decades has come to affect many species of storage insects and a wide range of insecticides, stems from careless use.

Resistance to phosphine fumigants in particular is almost certainly due in part to the not infrequent use of grossly inefficient application techniques. There is general agreement that the rate of resistance development, in any particular pest species and to any particular pesticide, is to some extent susceptible to control (i.e. management). Possibilities for the containment of resistance include sustained improvements in application techniques, where acceptable pesticide dosage rates remain effective, and the adoption of alternative pesticides or other control measures where the degree of resistance to a particular pesticide prohibits its use.

Continued monitoring of resistance, in field populations, is also necessary. Concepts of reducing resistance by genetic control have progressed to some extent and are not implausible. However, their practicality and cost effectiveness in stored-grains pest control remain undemonstrated. Moreover, it has been shown that some resistant strains of several storage insects have a reduced capacity to cause grain damage, through an impaired population growth rate. There is no reason to suppose that all resistant biotypes will be less damaging than their susceptible counterparts but the monitoring of this capacity, in field populations, is feasible and seems worthwhile.

Conversely, it would seem illogical to foster a reversion to susceptibility in cases where the resistant biotype has a substantially reduced capacity to cause damage and, therefore, a considerably reduced pest status. The possibility that insects may develop resistance to other control measures has been noted already with regard to reports of increased insect tolerance to low oxygen tensions in controlled atmosphere storage. As in the development of pesticide resistance, considerable genetic flexibility is to be expected. However, one may also expect that insect adaptability to massive constraints on fundamental biotic requirements, such as aerobic conditions, will be more limited than their ability to exploit the more natural genetic potential for metabolising extraneous toxicants.

10

Insect Management in Stored-Grain

Many different insects inhabit stored grain. Three major groups of stored-grain insects can usefully be identified: primary (or internal) pests, "secondary" beetles, and surface-feeding larvae. These pest categories are briefly described below.

Primary Pests

The most damaging pests of stored grain are those that develop within grain kernels. Adults deposit eggs on, or in whole kernels, and larvae develop hidden within the kernels. Damage caused by internal pests makes grain more susceptible to infestation by insects that feed externally on grain or grain debris. The common primary pests of grains in Nebraska are the weevils-rice weevil, maize weevil, and granary weevil. Other primary pests include the lesser grain borer and the Angoumois grain moth. Although inspectors may designate grain as "weevily" when in fact other insects are actually present, weevils are a unique and recognisable group of stored-grain insects. Adults are small (between 1/16 and 1/8 inch in length), but identifiable because the head has a prolonged snout.

Once the weevil larva completes development, the adult emerges from the damaged kernel and feeds, mates, and deposits eggs as it moves about among kernels. Another primary pest found in wheat, but rarely in corn, is the lesser grain borer. Identification of this pest is possible because the adult lesser grain borer's head projects downward, not forward, from the front section of the body. These insects may be found in any part of a grain mass; they are not restricted to the grain surface. Larval stages of the Angoumois grain moth also feed within grain kernels. This insect can infest grain in the field; storage infestations usually are limited to the surface layer of a grain mass, or in crib storage of ear corn.

Secondary Beetles

A second group of stored-grain insects includes the secondary beetles that develop and feed outside grain kernels or within cracked or damaged kernels. Most insects commonly collected in stored grain in Nebraska are beetles that range from /16- to -inch in length. Adults of most species are reddish brown to black, and their forewings are hardened to form a "shell" over the body. Larvae of common species are cream coloured and cylindrical. Species often collected in Nebraska include the saw toothed grain beetle, flat grain beetle, rusty grain beetle, foreign grain beetle, hairy fungus beetle, larger black flour beetle, and red flour beetle.

Like the weevils, these beetles are not limited in distribution to the grain surface, but instead inhabit any part of a grain mass. They feed on several different grains, but their buildup in any grain depends on broken kernels (or other fine material) or fungal growth on moist grain. This dependence on already-damaged grain explains the description of these insects as secondary pests, "bran bugs;" or fungus feeders. Concentrations of stored-grain beetles raise grain temperature and moisture, and these altered conditions favour continued population growth.

Surface-Feeding Larvae

The third major group of stored-grain pests is comprised of surface-feeding larvae, of which the Indian meal moth is the best known in Nebraska. These insects commonly inhabit the outer parts of a grain mass. Larvae reach approximately -inch in length and are cream coloured. They produce fine silky webbing as they feed and move about; mature larvae pupate within silken cocoons. Adult moths fly and mate in the bin headspace, where they may be seen resting on the bin walls and roof.

The Indian meal moth has a wingspan of about -inch; the outer portion of each forewing is reddish brown or copper coloured. This insect often presents a special management problem, because many populations in Nebraska have developed resistance to the insecticides malathion and dipel. Additional pests that sometimes infest stored grains in Nebraska include psocids (booklice) and grain mites. These soft-bodied pests feed on grain-rotting fungi.

An abundance of psocids or grain mites often indicates a more important problem of mold-related deterioration of the grain. Not all insects in grain are pests. Parasitic wasps, predaceous fly larvae, and predaceous Hemiptera (true bugs) attack certain grain pests. In addition, many field insects are inadvertently transported to grain bins where they cause no damage.

The behavioural and biological characteristics of the groups of stored-grain insects described above are important to management decisions. For example, an insecticide treatment incorporated in the surface part of a grain mass may provide adequate control of the Indian meal moth, but such an application will not control a secondary beetle or weevil problem that already exists in the core of the grain mass. Cleaning grain to remove fine material will greatly contribute to controlling infestations of several secondary beetles, but this action is of only limited value for reducing infestations of weevils or other primary pests. Although some stored-grain insects may infest maturing grain crops in the field, the extent of field-originated storage problems is minor in Nebraska.

Infestations more commonly develop when insects move to new grain from carry-over grain, from small amounts of grain not cleaned from empty bins, from feed supply buildings, and from grain debris beneath perforated floors of bins. Most pest species can fly at least short distances to reach new grain. Pest migration to, and development within, stored grain depends on suitable air and grain temperatures. Common drying and aeration practices usually produce moisture and temperature conditions that allow a fall-harvested crop such as corn to remain nearly insect-free during the first winter of storage.

As air and grain temperatures increase during spring and summer, insects move into stored corn and reproduce rapidly. When corn remains in storage through additional seasons, normal winter cooling causes some insect mortality, but many individuals, although inactive, do survive. Insect density usually increases with each year of storage. The level of insect buildup for any duration of storage varies among bins according to grain conditions (moisture, temperature, amount of fine material and mold), proximity of pest sources, and the use of insecticides.

Some generalisations provide an overall understanding of the population dynamics of these insects. First, reproductive potential is phenomenal; females of the various species can produce between 50 and 500 eggs each. In optimum conditions, individuals of several major species develop from egg to adult in approximately 4 to 6 weeks; but a few species are characterised by the development of only one generation per year.

Although temperatures for optimum development differ among species, the majority develop most rapidly and successfully at temperatures between 75 and 90F. At temperatures below 50F, stored-grain insects exhibit little or no activity. Cooling grain to temperatures below 50F for an extended period causes some mortality, and the percent mortality is greatest at even lower temperatures. Some pest species can, however, survive extended exposure to temperatures at or below freezing. Grain moisture also influences species composition and population density.

Several pests, including foreign grain beetle, hairy fungus beetle, and the meal worms, feed on fungi that develop in grain that is too wet. Others, such as the weevils and the larger black flour beetle, are better able to feed directly on grain that is high in moisture. Storage at moisture levels of 12 to 13 percent reduces insect proliferation, but some species will invade grain held at even such low moistures.

Insect Damage Grains

Insects damage stored grain directly by consuming kernels and reducing the weight and feed value of the infested commodity. Additionally, insects infesting stored grain produce heat and moisture that contribute to mold growth and spoilage. Owners of infested grains first suffer the losses that occur before sale; and then upon attempts to deliver grain, face discounts or quarantines if live insects are detected in grain samples. When discounts are assigned for insect infestation, insect numbers and type of grain are important factors.

Insect-damaged kernels also cause grade reduction and price discount. To manage stored-grain insects effectively, operators of grain storage facilities must examine grain for insect infestation before it is unloaded and moved into storage, and regular inspections must be repeated throughout the subsequent storage period. Initial inspection may use manual or hydraulic probes that withdraw grain samples from incoming loads. These samples should be screened using an appropriate grain sieve, and the screenings (material that passes through the sieve) should be examined for the presence of insects.

The size and shape of sieve openings vary for different grains. If the grain temperature is less than 60F, allowing the screenings to warm to room temperature before examination will activate any insects that are present; their movement increases the likelihood that an inspector can detect their presence. Infested loads should not be moved into storage with uninfested grain; doing so promotes infestation of the entire storage.

Once grain has been stored, regular inspections are necessary to measure grain temperature and moisture and to detect any insect infestations. The number of samples necessary to determine grain condition adequately is not well defined. The minimum number of samples suggested for round bins is given in Table 1, but increasing the number of sample sites increases the accuracy of the information obtained.

If temperature measurements are unusually high at any sites, samples should be collected from these areas to determine moisture content and presence of insects. It also is especially important to check for insects in samples from the grain surface and from areas where fine material is concentrated. Infestations often begin in these locations.

Table 1. Minimum Number of Samples for Determining Temperature, Moisture, and Insect Levels in Round Bins

	Temperature probes		Moisture and insect determination	
Bin diameter	Shallow	Deep	Shallow	Deep
Less than 24 feet	2	3	2	5
More than 24 feet	2	5	3	10

Shallow samples should be taken at or just below the surface of the grain mass. Deep samples should be taken from various depths determined by sampling equipment and ability to probe the grain mass. One shallow sample and one deep sample should be taken from the centre of the grain mass, especially if fine material is concentrated in the central core.

One standard sampling tool used to collect grain samples is the partitioned (or compartmentalised) grain trier. It is inserted into grain at about a 10 angle from vertical, with the sampling compartments closed and facing slightly upward. Rotating the handle opens the compartments; and three quick, short, vertical strokes force grain into each compartment. After the trier is filled, the compartments are closed and the trier is removed from the grain. A grain canvas or eaves trough can be used to catch the sample when the trier is opened. Triers can also be used to collect surface samples by inserting the trier horizontally just below the grain surface with the compartments opening upward. Three-five-, ten-, and twelve-foot triers are marketed by grain industry suppliers.

Deep-bin or deep-cup probes can be used to collect samples from greater depths within a grain mass. The sample cup is attached to the end of a metal probe and inserted closed into the grain mass. Extension sections added to the top of the metal probe allow insertion of the sample cup to the desired depth. Pulling up on the probe opens the sample cup, and it fills with grain.

If grain has settled within a storage, it is usually very difficult to insert a deep-bin cup beyond 10 to 15 feet below the surface. Collecting samples from depths greater than 10 to 15 feet requires a vacuum sampler or power probe. A shop vacuum can be modified to collect samples from depths as great as 20 to 30 feet. Hydraulic or mechanically powered probes can collect samples from depths as great as 100 feet.

Traps

Several types of traps are available to monitor stored-grain insects. A plastic probe or pitfall trap can be inserted into stored grain. This trap consists of a 14-inch long, 1-inch diameter clear plastic cylinder with perforations drilled through the upper part. The lower part of the cylinder is not perforated and contains a smaller plastic catch-tube capped with an open funnel. The top of the 14 inch cylinder is capped; the bottom is closed by a removable, tapered plug.

Insects crawl through the perforations into the trap, drop through the funnel into the lower catch-tube, and cannot escape. Although specific sex attractants, aggregation attractants, or food attractants can be used to lure insects into this trap, the random movement of insects within an infested grain mass produces substantial captures even in unbaited traps.

Plastic probe or pitfall traps should be placed just below the grain surface or probed into grain to a depth of about 12 feet. To probe the traps into grain, a metal collar is screwed onto a probe extension section; this collar holds the trap in a vertical position as it is pushed into the grain. For retrieval of the trap from the grain mass, the top of the trap contains openings for the attachment of a rope. Pitfall traps should be left in place for one to four days. Increasing the trapping period increases the likelihood that insects will be captured. Because these traps depend on insect movement, they are not effective in grain colder than approximately 50F where insects are inactive.

Additional traps designed for monitoring stored-grain insects include corrugated cardboard traps that hold an oil lure that attracts and kills insects active where the trap is placed. These traps might be used on the surface of a grain mass to detect insect presence. They are also effective around bagged seeds or feeds in warehouses. Paper sticky traps baited with attractants are also available.

In Nebraska the most commonly used of these traps contains the Indian meal moth pheromone, and it is used to detect the presence of this pest in or around stored grain or grain products. Paper sticky traps are also used to monitor flight activity of the lesser grain borer, cigarette beetle, and several other pests in warehouses and processing plants.

Methods for Controlling

Controlling established infestations of stored-grain insects is often difficult. Usually, preventing infestations is more profitable and successful than eliminating problems that have already developed. Preventive steps include thorough sanitation, application of an insecticide to the empty storage, proper cleaning and drying of grain, application of protectant insecticides when extended storage is planned, and adequate aeration for temperature and moisture management.

An important rule for successful grain storage is that new grain should never be stored atop old, carry-over grain. When this rule is ignored, insects that infest old grain will readily move into the new grain placed in the same storage. Preventing insect carry-over within a bin is also the reason for thoroughly cleaning bins before adding new grain. Cleanup should include removal of all grain that may be caked or webbed on the bin walls and all grain and debris on the bin floor.

Persons doing cleanup work inside a dusty storage facility should always wear a dust-filtering mask. Although removing the perforated floor is impractical in most bins, cleanup practices that include the subfloor plenum are recommended where possible. Also remove grain and grain debris from combines, wagons, augers, etc. Piles of spilled grain near bins also are sources for insect infestations; these infestation sources should be removed before new grain is moved into storage.

Bin Sprays and Empty Bin Fumigation

Applying a registered insecticide to the walls and floor of empty bins supplements, but does not replace, cleanup efforts. Insecticide residues control insects that may have remained in hard-clean cracks and crevices or beneath the perforated floor. Sprays should be applied to the point of run-off; applicators should thoroughly treat all cracks and crevices and around doors.

Applicators applying empty-bin sprays inside storage structures should wear a cartridge or canister respirator to prevent inhalation of insecticide vapours. Directing extra spray to and through perforated flooring will provide some control of insects in the subfloor plenum, but maximum control of insects in this space requires fumigation or the removal of the perforated floor and thorough cleanup. Fumigating empty bins to control insects in the subfloor plenum may be necessary if grain will be held in storage for more than one month during warm weather from June through October.

Moist Grain

Whenever plans include storing grain during summer months, drying to moisture levels of 12 to 13 percent is recommended. This range is drier than the preferred condition for stored-grain insects, and it slows their buildup. In addition, insecticides applied directly to grain persist longer on dry grain than on moist grain.

Fine material (in corn, called BCFM, or broken corn and foreign material) in stored grain is detrimental for several reasons. Broken kernels, weed seeds, and other crop debris often spoil at moisture levels generally considered safe for whole kernel storage. Concentrations of fine material reduce air flow and prevent uniform aeration of a grain mass. Extremely fine particles form aerosol dusts that can explode. In addition to these negative effects in grain, fine material also contributes directly or indirectly to most insect problems in stored grain.

Many secondary beetles depend on fine material to survive and reproduce. These insects release metabolic heat and moisture that contribute to the survival and reproduction of additional secondary pests and weevils.

Cleaning grain to remove fine material reduces the survival of the most common pests found in grain in Nebraska. Rotary cleaners and aspirators are more effective than perforated or screened sections of augers.

The following are some important general considerations for insecticide use:

- Methoxychlor and cyfluthrin (Tempo®) are approved only as empty-bin sprays; they may be applied to bin walls and flooring before filling the bin, but they must not to be applied directly to grain.
- Most of the products mentioned above may not be applied to soybeans; only Bacillus thuringiensis, diatomaceous earth, and certain fumigants may be applied directly to soybeans.
- Many populations of Indian meal moth are resistant to malathion, so where this insecticide is used to control beetles and weevils, additional treatments are necessary for Indian meal moth control at, and just below, the surface of a grain mass.
- Bacillus thuringiensis is a bacterium that acts as a stomach poison. It must be ingested to be effective. It controls only the larval stage of butterflies and moths, and its most common use is for Indian meal moth control.
- Diatomaceous earth is a chemically inert (inactive) but physically abrasive dust that damages an insect's cuticle (body covering). It must be applied at fairly high rates to control stored-grain insects.
- Pyrethrins provide rapid knockdown of insects in the headspace of a storage. They break down rapidly and are not effective for residual control.

Products containing pyrethrins should include a synergist (often piperonyl butoxide) to be effective.

- Methoprene was registered in 1988 for use on stored grains, but not on soybeans. Methoprene is similar to the naturally occurring hormones of insects, and it interferes with the growth and maturation of immature insects. It will not control adults already present in grain, but it will prevent immature stages from developing to adults and reproducing.
- Reldan® and Actellic® are effective against Indian meal moth and the weevils and other beetles that infest grain. These products are registered for use on only certain grains.
- When Bacillus thuringiensis, chlorpyrifos-methyl (Reldan®), diatomaceous earth, malathion, methoprene, pirimiphos-methyl (Actellic®), or pyrethrins are applied directly to grain at rates listed on the product labels, there is no required waiting period for subsequent sale, processing, or feeding. Treated grains can be used immediately.
- Manufacturers of all pesticides must produce a material safety data sheet (MSDS) that describes the physical and chemical hazards of each pesticide product. Users should have a current MSDS) available whenever they handle a pesticide.

It is important to remember that protectant insecticides are not fumigants. With the exception of Bacillus thuringiensis, stored-grain insecticides, whether applied to prevent infestation or to control an existing problem, are contact poisons. Residues on treated grains may provide insect control for over one year. Their activity, however, is limited to the grain on which they are directly applied. These insecticides do not readily form a gas and will not kill insects present in untreated portions of the grain mass.

Fumigants act as gases that spread throughout the closed space into which they are released. They dissipate readily; and no effective insecticidal residue remains following aeration several days after application. The stability of stored grain insecticides and the duration of their effectiveness are determined by grain temperature and moisture. Temperatures below 40F not only prevent insect activity, they also retard the degradation of insecticides.

At 60F and 12 percent moisture, a malathion residue of 6.2 parts per million (ppm) is present one year after application of the labelled rate of 10 parts per million. A 3 parts per million residue is considered effective in the absence of resistance. At 14 percent moisture and 60F, a residue of only 2.4 parts per million remains one year after application at 10 parts per million. This moisture- and temperature-related breakdown of malathion explains why storage at low temperatures and 12 to 13 percent moisture is recommended when long-term efficacy of malathion is needed.

Although information about Actellic® and Reldan® is less abundant than for malathion, both compounds are somewhat more persistent than malathion. Actellic® residues on treated wheat are very similar for storage at 12 and 16 percent moisture levels. Although the moisture dependency shown for malathion seems less of a concern for Actellic® and Reldan®, these compounds have not been used as extensively as malathion, and the extent to which they may outperform malathion is still somewhat unclear.

Application Methods

Grain can be treated in several ways as it is augured into

storage or transferred from bin to bin. One simple and effective device is the gravity flow applicator, which delivers an insecticide solution to grain as it flows into an auger, conveyor belt, or similar grain transfer equipment. It requires no electrical or mechanical power source. Its successful use does, however, require knowledge of the delivery rate of the auger, a value easily computed by determining the time required to move a known quantity of grain into storage.

Stored-grain insecticides are effective when applied in 5 gallons of water per 1,000 bushels of grain. Spray and dust applicators are commercially available; grain handlers may also choose to custom-design their own equipment. Spray-delivered under constant pressure may provide a more uniform delivery rate than gravity flow applicators. The uniformity of coverage (that is, whether each kernel is coated) required for effective insecticide activity allows for some flexibility in application methods. Research has shown that it is not necessary to obtain uniform coverage of all kernels; it is important that an insecticide be applied at a consistent rate throughout all grain layers.

To deliver the desired insecticide rate, it is usually best to apply the treatment as near as practical to the grain's destination; this minimises the amount of insecticide left on grain moving equipment add maximises the deposit remaining on the grain. A coarse spray application such as that provided by a flooding nozzle produces maximum deposits on grain and minimum losses to aerosol grain dusts.

Stored Grain Management without Fumigation

When insects are detected in stored grain, several possible management practices should be considered.

When insect infestations are detected early and pest densities are very low, the most profitable action may be to clean and move the grain immediately without any chemical treatment. This option may be especially appropriate where traps provide a very sensitive detection programme and infestations consist of secondary beetles; moving and cleaning grain can eliminate a high percentage of these beetles.

During cool weather, aerating to cool the grain below 50F can prevent insect activity and allow an extended period of safe storage. Further cooling provides some degree of control, but some stages of most pests do survive the cooling procedures usually applied to stored grains. Once grain has been cooled, fumigation may be ineffective or prohibited until the grain has been rewarmed. Sometimes insect infestations are limited primarily to the surface and central core of a grain mass.

Light infestations of Indian meal moth near the surface of the grain can be controlled by incorporating registered insecticides other than malathion. None of the stored-grain insecticides applied as a surface treatment without incorporation will control Indian meal moth larvae already present a few inches below the grain surface; incorporation by raking is necessary. Abundant webbing should be raked from the surface before treating.

Fumigation may be necessary in these situations. Where secondary beetles are mainly confined to a central core with concentrations of fine material, removing that central core can reduce the risk of severe damage in the remaining grain. The grain removed from the core can be cleaned, treated with a registered insecticide, and returned to the same bin. Remaining grain containing less fine material will not be insect-free, but additional management and successful storage of this grain will be easier once the excess fine material in the centre of the storage has been removed.

Where infested grain can be moved to a clean, empty storage, transfer and treatment with a protectant insecticide is recommended. Grain also should be cleaned during transfer. Residual insecticides will not immediately kill the immature stages of insects that are developing within grain kernels, but insecticide residues will eventually control the adult insects that emerge from the kernels. Residues will persist on grain and prevent reinfestation for an extended period, the length of which depends on grain moisture and temperature.

Fumigation in Grain Storages

Effective fumigations result from two factors: (1) adequate and uniform fumigant concentration and (2) adequate fumigation time. These conditions are provided by thoroughly sealing the storage, correctly applying the proper amount of fumigant, and leaving the storage sealed for the label-specified period of time. Over-applying a fumigant will not compensate for failing to seal a storage, ailing to distribute the fumigant properly, or failing to keep the storage sealed for the necessary length of time. If application rate, fumigant distribution, sealing efforts, or fumigation time is inadequate, the degree of insect control will be reduced.

Aluminum Phosphide

Aluminum phosphide pellets and tablets can be introduced into grain by adding them to the grain stream as grain is moved into a storage or by probing them into an existing grain mass. There are advantages and problems associated with each application method. Commercially available automatic dispensers meter the application of pellets or tablets to moving grain carried by augers, conveyor belts, or bucket elevators (Figure 1).

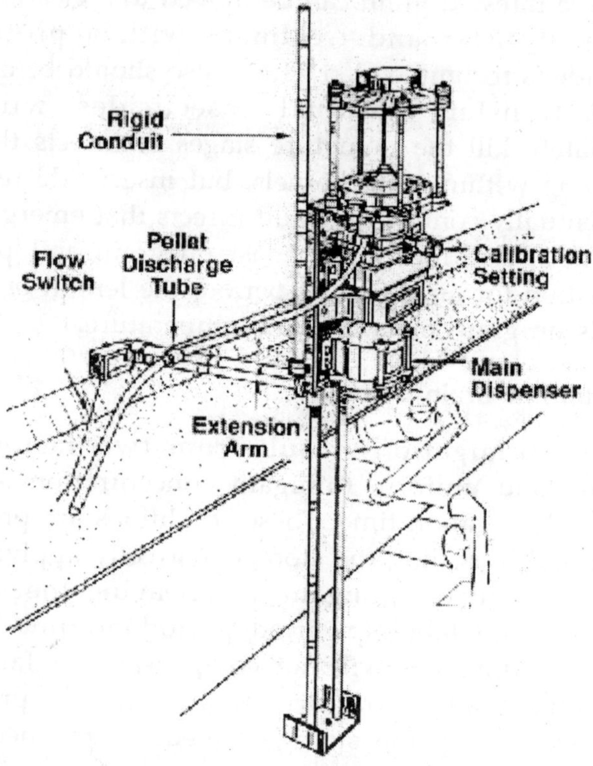

Figure 1. Automatic metering device for aluminum phosphide tablets and pellets.

The main benefit of applying aluminum phosphide to a moving grain stream is that it allows a very uniform distribution of pellets or tablets within the grain mass as the storage fills. This application method does not require that workers enter the actual fumigation atmosphere, and the difficulties associated with probing aluminum phosphide into the grain mass are avoided. Two trained applicators need not be present when aluminum phosphide is applied in this way; outdoor application of aluminum phosphide to a moving grain stream requires only one certified applicator. Problems that complicate this type of application are related to gas release and escape during the filling process.

Gas evolution from tablets or pellets normally begins within one to two hours of application, and sometimes much faster. Some gas obviously escapes from the storage as grain is continuously added because the storage is not sealed during filling. This problem may be insignificant where bins or tanks can be filled continuously and rapidly; but when filling is interrupted, gas escape means insufficient concentrations near the top of the grain mass. As a general rule, applying aluminum phosphide in the grain stream as a storage is filled is appropriate only if the storage can be filled and closed within one work day. Where aluminum phosphide is metered into the grain stream, it is important to monitor gas concentrations in confined work areas where hydrogen phosphide may accumulate.

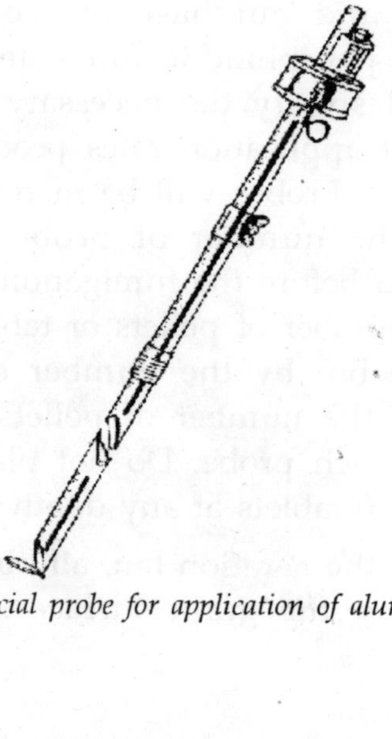

Figure 2. Commercial probe for application of aluminum phosphide.

Most insect infestations that must be managed by fumigation involve grain that cannot be moved from one bin to another to allow metered application of aluminum phosphide to the grain stream. In these instances, probing tablets or pellets into the grain mass is usually necessary. Special probes for aluminum phosphide are commercially available (Figure 2); applicators also may use probes constructed from plastic PVC pipe. Sprinkling pellets or tablets onto the grain surface or raking them into the top layer of grain will not provide an adequate fumigation of grain below the surface.

Small bin fumigation

Application methods may differ according to the type of storage. In most round bins with capacities less than 25,000 bushels, two trained people can effectively apply aluminum phosphide. Labels and good sense require that two people always be present during a fumigation. Steps for fumigating bins that hold less than 25,000 bushels of grain include:

1. Calculate and purchase the correct amount of aluminum phosphide to fumigate the bin. Product labels will specify the necessary application rate; insufficient application rates produce unsuccessful fumigations. Probes will be made on a 5-foot grid pattern; the number of probe sites should be determined before the fumigation is begun. Divide the total number of pellets or tablets to be used in the entire bin by the number of probe sites to determine the number of pellets or tablets to be placed in each probe. Do not place more than 50 pellets or 20 tablets at any depth in a single probe.
2. Except for the aeration fan, all bin openings below the level of the grain surface should be sealed

before the fumigation is started. Use 4-mil (0.04 inch) polyethylene sheeting and duct tape to cover the side door, unloading auger shaft, and any other openings that might allow gas leakage.

3. Precut a piece of polyethylene sheeting to cover the surface of the grain. Make it large enough to cover any peak (a better approach is to level off the peak) and allow for an extra 2 feet to tuck along the walls. Also attach a rope or cord to the polyethylene cover so that it can be pulled from the surface of the grain without entering the bin.
4. Plan all fumigation steps before making the first probe. Working efficiently, application can be completed in less than one-half hour, and problems of heat exhaustion and buildup of hydrogen phosphide concentrations in the bin headspace can be avoided.
5. One trained applicator should begin by tucking the polyethylene sheeting along the wall opposite the entrance hatch. The other trained applicator should begin probing pellets or tablets at 5-foot horizontal intervals. The probe should be inserted to its full length (at least 5 feet), and pellets or tablets dropped into the probe as it is withdrawn. Hydrogen phosphide gas will not settle to provide adequate and uniform concentrations more than 30 feet below its application site, so surface probes will not provide adequate fumigation of taller structures. Applicators handling aluminum phosphide should wear dry cotton gloves. The applicator not making the probes should periodically use a gas detector to monitor hydrogen phosphide concentrations during the application.

Detectors manufactured by Auer, Draeger, Matheson-Kitagawa, MSA, and Sensidyne can be used. If gas concentrations exceed 0.3 parts per million, but not 15 parts per million, a full-face, NIOSH/MSHA-approved canister-type gas mask (or canister respirator) must be worn by any person inside the bin. Canisters used with these respirators should be approved for effectiveness in filtering hydrogen phosphide. A canister respirator (or canister gas mask) or a self-contained breathing apparatus (SCBA) must be available within the bin even if gas concentrations are less than 0.3 parts per million. If hydrogen phosphide concentrations exceed 15 parts per million, a NIOSH/MSHA-approved SCBA must be worn. Even if measured concentrations of hydrogen phosphide remain below 15 parts per million during application, a SCBA must be available at the site or locally at a fire department, rescue squad, or similar source. Applicators must be trained in the use of a SCBA.

6. After aluminum phosphide has been applied at the last probe site, the polyethylene sheeting should be tucked along the remainder of the wall. The polyethylene cover will not prevent diffusion of hydrogen phosphide into the headspace of the storage; but it will prevent air currents from actively carrying the fumigant up and out of the bin. If the grain surface is not covered, the fumigant rate must be increased to provide adequate fumigant concentration in the bin headspace as well as within the grain mass. Eaves and roof hatches must be tightly sealed if the grain surface is not covered.
7. Aluminum phosphide can also be introduced into the subfloor plenum through the aeration or drying

fan opening. Use a PVC pipe probe to place pellets or tablets as far as possible from the fan opening. This application supplements, but does not replace recommended probe applications. Do not pile pellets' or tablets at a single location in the subfloor plenum. Do not place aluminum phosphide into the subfloor plenum if there is any chance that water has seeped or leaked onto the concrete surface. If aluminum phosphide contacts water, an explosion is likely. The opening to the aeration duct or fan should be covered with polyethylene sheeting and sealed with duct tape once the application is completed. The aeration fan should be left off during the entire fumigation.

8. Post warning placards on all doors and near the ladder. These placards must bear the words DANGER/PELIGRO and the skull and crossbones symbol. Placards must identify the date of fumigation, the name of the fumigant applied, and the name, address, and telephone number of the applicator. Each placard must state "AREA UNDER FUMIGATION, DO NOT ENTER/NO ENTRE."
9. Fumigated bins should remain sealed for the period specified on the fumigant label; this period ranges from 3 to 10 days depending on grain temperature. Colder temperatures require periods longer than are necessary for warm grain. Labels prohibit fumigation when grain temperatures are less than 40F; highly effective fumigations are difficult to accomplish when grain temperatures drop below 60 to 65F. Following the 3- to 10-day exposure period, remove the polyethylene sheeting from the grain surface and aeration fans and open ventilation hatches. Aerate the grain until hydrogen phosphide

levels drop below 0.3 parts per million; a gas detector must be used to determine fumigant concentrations in the bin headspace. Placards may be removed and unprotected personnel may be allowed to enter the bin only after a certified applicator has used an appropriate gas detection device to determine fumigant concentration.

10. Following aeration, a fumigated bin is susceptible to reinfestation. Surface topdress applications of protectant insecticides can limit reinfestation.

Flat storage fumigation

Fumigating a flat storage with aluminum phosphide requires application methods and concerns similar to those outlined for small bins. Nonetheless, different problems are encountered. In some ways, a flat storage fumigation resembles a warehouse fumigation more than a bin-type fumigation. Prefumigation planning should be completed in much the same way as described previously. Total fumigant needs and the number of pellets or tablets per probe should be determined before the application is started. It is often impractical to cover the grain surface in flat storages; and where grain remains uncovered, the calculation of how much fumigant to use must account for fumigating the headspace as well as the grain.

Fumigating large flat storages requires a great number of man-hours of labour; four or more applicators are usually necessary for probing and applying aluminum phosphide, while two or three additional trained personnel assist with opening and collecting flasks and monitoring gas concentrations. Vents and openings in flat storages are usually sealed after the application of aluminum phosphide pellets or tablets. Leaving vents open during application slows the buildup of gas concentrations in the headspace during the lengthy application process.

Concentrations of hydrogen phosphide gas commonly exceed 0.3 parts per million in the headspace of flat storages during application. Monitoring fumigant concentrations is essential, and full-face canister respirators and SCBAs must be available for all applicators. These canister respirators, containing fresh new canisters approved for aluminum phosphide, must be worn in hydrogen phosphide concentrations of 0.3 to 15 parts per million. A SCBA must be worn if hydrogen phosphide concentrations exceed 15 parts per million.

A SCBA will supply up to 25 to 30 minutes of air; but during hard Work in a hot storage, applicators may consume the air supply more rapidly. Once an application is completed, vents and hatches should be sealed from the outside. Duct tape or a similar product should be used to tightly seal 4-mil (0.004 inch) polyethylene sheeting around vents and doors. Up to 20 percent of the aluminum phosphide pellets or tablets required to fumigate the entire storage can be applied in the aeration ducts of flat storages. This may be especially appropriate where numerous aeration ducts lead into the storage from side walls.

Applications should be made as described for small round bins, and the same cautions also apply about standing water and the maximum number of pellets or tablets to place at one location. Once applications are made at the aeration ducts, polyethylene sheeting should be taped in place to seal duct openings or fans. The regulations and safety concerns discussed for small bin fumigation apply as well for flat storage fumigation with aluminum phosphide.

Grain temperature must be at least 40F for fumigation to be legal; and warmer temperatures increase the likelihood of successful control. The use of gas detection devices and respiratory protection equipment may be even more important in flat storage fumigation, because application time and potential exposure to phosphine gas often are greater.

Fumigating flat storages is often less successful than expected. Limited success results from inadequately sealing flat storages. Thorough sealing is often difficult and impractical. Few applicators attempt to seal around the eaves of flat storages, but openings at the eaves can allow rapid escape of the fumigant. Immediately following fumigation, insect mortality may seem very high. This success is only apparent because fumigant concentrations high enough to kill adult insects exposed between grain kernels did develop within the bin, and initial inspections therefore reveal dead adult insects.

However, in poorly sealed storages, fumigant concentrations may not reach necessary levels within kernels, in dense pockets of fine material, or near leaks at walls, eaves, and other openings. Where poor sealing produces sublethal concentrations of fumigant gas, immature insects within kernels and all stages of insects in poorly penetrated areas may survive fumigation. Because successful fumigation of flat storages can be difficult, thorough preventive measures are especially important when long-term storage in these structures is planned.

Fumigation in large bins and tanks

Large bins or storage tanks taller than 30 feet are not easily fumigated with aluminum phosphide except where the bin can be "cored." Because hydrogen phosphide gas does not readily settle to reach lower portions of these tall bins, tablets or pellets placed by surface probes are not deep enough for uniform distribution of the fumigant gas. This limitation can be overcome, at least partially, by unloading grain to extract the bin's central core and returning that grain into the top of the same bin with a metered application of aluminum phosphide.

Continuous unloading and refilling of the core causes aluminum phosphide-treated grain to be reintroduced to the top of the bin and drawn down to the bottom of the bin. Such cycling allows application of aluminum phosphide to a moving grain stream without transfer to an empty bin. Application of aluminum phosphide while grain is cycled within a single large bin requires prior calculation of the number and rate of tablets or pellets to administer. The goal is to have almost all of the correct total dosage applied by the time the first pellets or tablets reach the bottom of the bin.

Applicators must know the grain transfer rate and have an accurate estimate of the number of bushels that must be cycled before the first pellets or tablets will reach the bottom of the bin. Although this method can effectively fumigate large bins, it requires thorough knowledge of the grain handling system in use. Other approaches (for example, a recirculating application of methyl bromide) also may be practical.

Hydrogen phosphide, especially in high concentrations and when relative humidity is high, corrodes copper, brass, gold, and other metals. Switches, communication devices, computers, calculators, and electric motors contain sensitive components that can be destroyed if exposed to hydrogen phosphide. Corrodible materials should be removed from the area to be fumigated or partially protected by tightly sealing them in polyethylene. It is important to point out that the aluminum phosphide application methods described above may not be used when fumigating most processed foods.

Pellets and tablets (or their reacted products) must not contact any processed food except processed brewers' rice, malt, or corn grits. This restriction means that aluminum phosphide cannot be probed into other processed commodities for which its use is approved. For such commodities, pellets or tablets must be applied on trays or similar containers, or prepackaged application devices such as "strips" or "blankets" must be used. Prepackaged strips and blankets can also be used in some instances in raw grain, but uniform distribution of hydrogen phosphide is usually achieved only when most of the required rate of tablets or pellets is probed into a grain mass.

Methyl Bromide

All persons who work with methyl bromide must be aware of the hazards associated with this fumigant, should understand the use of gas detection devices and respiratory protection equipment, and should be trained in proper application methods and emergency procedures. When methyl bromide is used to fumigate enclosed spaces such as grain storage facilities, two trained persons must be present at all times when worker exposure exceeds 5 parts per million.

Applicators should avoid wearing tight clothing, jewellery, gloves, and boots when applying methyl bromide. Methyl bromide may be trapped next to the body under such clothing and can severely injury the skin. Any contaminated clothing or boots must be removed and not reused until thoroughly cleaned and aerated. Whenever full-face respiratory protection is not required, applicators should wear goggles or a full-face shield for eye protection when handling methyl bromide liquid.

If the concentration of methyl bromide in the work area, as measured by a gas detector tube and pump, does not exceed 5 parts per million, no respiratory protection is required. If this concentration is exceeded at any time, all persons in the fumigation area must wear a NIOSH/ MSHA-approved SCBA or combination air-supplied-SCBA respirator. Canister-type gas masks with air filtering cartridges are not adequate for use with methyl bromide. For recirculating application of methyl bromide for control of insects in stored grain, the following steps should be followed.

1. The storage should be sealed thoroughly, using tape for small openings and polyethylene sheeting secured with tape for large openings. Unsealed openings allow methyl bromide to escape during the recirculation application. Fumigant leaks reduce the effectiveness of the fumigation and can present sad concerns around the storage.
2. Applicators must post warning placards at all entrances to the fumigated storage. Placards must present the words DANGER/PELIGRO and the skull and crossbones symbol. Warnings also must state "AREA UNDER FUMIGATION, DO NOT ENTER/NO ENTRO." Placards should state the date of the fumigation, the name of the fumigant applied, and the name, address, and telephone number of the applicator. Any person who transfers a treated commodity to another site without aeration must ensure that the new site is placarded until the commodity is aerated sufficiently to

reduce methyl bromide concentrations to below 5 parts per million. Only a certified applicator may remove the placards, and only after one or more gas readings have indicated that the methyl bromide concentration is below 5 parts per million.

3. Methyl bromide should be applied at a rate of 1.5 to 6.0 pounds per 1,000 cubic feet. Follow label instructions to determine the rate for the specific commodity to be fumigated. A duct system should be constructed so that the aeration system can be used to force air through the grain mass and collect and recirculate that air back through the aeration fans. Methyl bromide can be discharged into the duct system or into the headspace of the storage.
4. For most stored grains, methyl bromide recirculation should continue for 24 hours; check label directions for fumigation periods for specific commodities. This period of 24 hours or less is one benefit associated with methyl bromide fumigation. Under most conditions, other fumigants require an exposure period of 72 hours or longer. During the application period, it is important to use gas detector tubes to check for methyl bromide leaks around the treated storage; leaks will reduce fumigation success and may cause health hazards in the work area.
5. Once the fumigation period has ended, storages must be aerated. To aerate, disconnect the return air flow from the recirculation system so that exhaust air discharges outside the storage. Aeration should continue until gas detection devices show methyl bromide levels have dropped below 5 parts per million. Multiple readings should be taken in the elevator headspace, access pits, and other low areas

to detect any possible pockets of methyl bromide gas.

6. As mentioned above, once fumigated grain is aerated, it is susceptible to reinfestation. Surface top-dress applications of protectant insecticides will help to limit reinfestation.

Methyl bromide must not be applied if the grain temperature is less than 40F, except in cold storage fumigations specified in APHIS quarantine programmes. Methyl bromide will retard or destroy the germination potential of seeds in many fumigation conditions. For grain to be used for seed, alternatives to methyl bromide fumigation are usually recommended.

Carbon Dioxide

Low concentrations of oxygen and increased concentrations of carbon dioxide (CO_2) or nitrogen (N_2) have been used to control insects for centuries. Storage of grains in hermetically sealed containers relies on the respiration of seeds, insects, or microorganisms to deplete oxygen supplies and kill insects present. In 1980 and 1981, the U.S. Environmental Protection Agency approved registrations for the use of CO_2, N_2, and combustion gases for the fumigation of a range of commodities, including stored grains. Because progress using atmospheric gases as fumigants has been greatest with CO_2, only CO_2 fumigation will be discussed. Carbon dioxide is a colourless, odorless, and tasteless gas that is approximately 1.5 times heavier than air.

It is noncombustible and, in fact, is used as a fire extinguishing chemical. Carbon dioxide is produced in large quantities, especially for the food processing industry, where it is used in freezing and refrigeration. It is commonly stored and transported as a liquid at 0F and under a pressure of 300 pounds per square inch (psi). Bulk containers range in capacity from 4 to 50 tons; smaller cylinders holding 20 to 100 pounds also are available. Carbon dioxide is a desirable fumigant because no toxic residues remain in treated grain. In addition, CO_2 does not alter the germination potential of treated grain, and it leaves no undesirable odors or flavours.

At high concentrations, CO_2 triggers increased respiration in insects (as well as in humans), and dehydration is one cause of death in insects exposed to high concentrations of CO_2. Most trials evaluating the effectiveness of CO_2 as a fumigant have utilised concentrations of approximately 60 percent. Concentrations substantially lower than 60 percent have been shown to be less effective; higher concentrations provide little or no added benefit. The effectiveness of 60 percent CO_2 as a fumigant is related to grain temperature, and duration of fumigation must be adjusted accordingly.

In grain held at temperatures less than 60F, CO_2 fumigation is generally ineffective because fumigation times become extremely long. As grain temperatures increase, fumigation times drop to 4 to 7 days at 80F. Table 2 summarises temperature-exposure time relationships for successful CO_2 fumigations.

Table 2. Recommended Exposure Times for Carbon Dioxide Fumigation

Commodity temperature	Days of exposure at 60% CO_2 atmosphere
60° F.	21-28
70° F.	10-14
80° F.	4-7

For successful fumigation, CO_2 must be introduced into a storage and displace existing air so that a 60 percent CO_2

concentration is reached. In a 20- to 30-thousand-bushel bin, the CO_2 application rate during this initial "purge" would commonly range from 300 to 600 pounds per hour. Following the initial purge of the storage atmosphere, additional CO_2 injections are made as necessary to maintain a concentration near 60 percent. In well-sealed concrete tanks, the total amount of CO_2 required for a 4-day fumigation (adequate if grain temperatures are approximately 80F) should range from 150 to 200 pounds per 1,000 bushels of grain. In intervented concrete silos, applicators will probably need to use 240 to 260 pounds of CO_2 per 1,000 bushels if the vents are not sealed.

In bolted steel bins, seams and bolt holes must be sealed to prevent excessive leakage and to allow successful fumigation using economical rates of CO_2. Estimates from 1989 indicate that the amount of CO_2 required for fumigation of warm grain (70 to 80F) should cost 0.5 to 4 cents per bushel, depending on sealing, leakage, and local CO_2 costs. Carbon dioxide can be injected into grain in storage by (1) a top-down purge in which CO_2 is added at the top of the structure and displaces air as it settles downward or (2) a lift in which CO, is added at the bottom of the structure.

The type of storage is an important factor in determining the better application method. The top-down purge is used in concrete elevators where the injection hose is placed in the top of a well-sealed tank. Bottom injection is usually best in stand-alone steel bins where major leaks occur at the top around eaves and hatches. In either application, the storage must be vented during the initial purge so that air pressure allows expulsion of the normal atmosphere. To determine when the purge has been completed, CO_2 concentrations should be measured at the grain surface or headspace (for bottom-applied "lifts") or at the base of the grain mass or aeration floor (for top-applied injections).

Following the initial purge, complete sealing produces the most successful fumigation. Additional injections of small amounts of CO_2 should be made when measurements indicate CO_2 concentrations within the commodity have dropped below 50 to 60 percent. Carbon dioxide application in grain storages requires a pressure regulator to drop pressures from 300 pounds per square inch in storage tanks to 10 to 40 pounds per square inch at the injection hose. A flowmeter and globe valve should be used to measure and set the flow of CO_2 to the desired rate, usually measured in pounds per hour.

A vaporiser is also required to speed the conversion of CO_2 from liquid to gas. Applicators must be able to monitor CO_2 at low concentrations to provide information for worker safety, and at high concentrations to determine the need for continued injection of CO_2 into the structure. Vacuum pumps and tubing should be used to draw gas samples from the structure during fumigation. Companies that produce detector tubes and pumps used in measuring concentrations of phosphine, methyl bromide, or chloropicrin also market tubes for measuring CO_2 concentrations.

Separate tubes (low and high range) must be used to detect concentrations that threaten worker safety (below 1 percent CO_2) and concentrations that provide adequate fumigation (about 60 percent). Thermal conductivity sensors capable of measuring CO_2 concentrations from 0 to 100 percent also are available; these cost $1,000 or more.

Although CO_2 fumigations leave no toxic residues in treated commodities, applicators must realise that CO_2 gas is poisonous at high concentrations in enclosed spaces. The concentration of CO_2 in the atmosphere is about 0.03 percent. The threshold limit value for CO_2 exposure is 1.0 percent. When controlled concentrations of CO_2 reach $_2$ percent, breathing rate increases by 50 percent above normal, and in 5 percent CO_2, human breathing is three times more rapid than normal. The results include strain, fatigue, and exhaustion. High concentrations not only reduce oxygen availability, but also trigger additional respiratory stress.

It is essential that CO_2 concentrations be monitored when applicators enter the fumigated structure and in enclosed work areas adjacent to the structure. Canister respirators and other equipment that filter a poisoned atmosphere do not provide protection in high-CO_2 and low-oxygen atmospheres. Only a SCBA provides adequate protection. Structures fumigated with CO_2 must be aerated before unprotected workers may enter. Posting warning placards and taking gas readings is recommended where CO_2 is used.

11

Fumigation Technology

The earliest use of fumigants dates back to about 1000 years B.C. when "sulphur fumes" were used to disinfect ancient homes. However, it was only in 1854 that the use of carbon disulphide and hydrogen cyanide for disinfestation was acknowledged. The advent of these chemicals gave rise to the modern concept of fumigation. To date, fumigation offers one of the most effective control measures against varieties of stored product pests.

Fumigation in Developing Countries

Fumigants which are applied in the gaseous state, such as hydrogen cyanide or methyl bromide, have used for many years in the developed countries. Certain fumigants applied as volatile liquids, e.g. ethylene dichloride, ethylene dibromide, carbon tetrachloride and carbon disulphide, have also been commonly used as grain fumigants, often as mixtures. It was only about 30 years ago that fumigation, mostly with methyl bromide or ethylene dichloride/carbon tetrachloride, began to be carried out in developing countries.

The introduction of phosphine, generated by exposing solid formulations containing aluminium or magnesium phosphide to air, greatly simplified application procedures so enabling fumigation to be used more readily in developing countries. Unfortunately the expansion in the use of phosphine was not always accompanied by appropriate training in the techniques of application needed to ensure a total kill of the insect infestation. Recently there has been increasing evidence showing the development of resistance by certain major pests to phosphine.

Levels of resistance in some strains of certain species have increased to such an extent that it is difficult to achieve successful results using a recommended dosage rate and fumigation procedure. There is much evidence to suggest that such resistance is associated with the use of poor fumigation techniques over many years. These difficulties with phosphine have been accompanied by a gradual withdrawal of the older alternatives which are either much less convenient to use or are now suspected of having carcinogenic properties.

The scope for the introduction of completely new fumigants which might be used as phosphine replacements is necessarily very limited due to the small number of chemicals possessing the necessary properties, and the high cost of producing all pre-registration data now required by pesticide registration authorities. The role of phosphine as a fumigant is crucial to developing countries and if its continued effectiveness were to become doubtful, without the existence of suitable alternatives a grave situation would be created.

Application of Phosphine

Phosphine is applied to commodities as a solid formulation containing aluminium or magnesium phosphide which reacts with atmospheric moisture to liberate phosphine. Formulations based on magnesium phosphide generate phosphine more rapidly than those containing aluminium phosphide. Phosphine is also liberated more rapidly from both types of formulation under ambient conditions of high relative humidity and temperature. The manufactured products contain ingredients which retard this reaction or produce carbon dioxide which prevents spontaneous combustion of phosphine.

Four main types of formulation are currently available for the generation of phosphine;

(1) flat or round tablets generating 1 g of phosphine,

(2) flat of round pellets generating 0.2 g of phosphine,

(3) sachets generating 11 g of phosphine and

(4) 'plates' generating 33 g of phosphine.

Of these only the fiat tablets and pellets are manufactured in developing countries. Each of these has particular characteristics in relation to the rate of generation of gas. Pest control organisations in developing countries need to consider which type is most suitable for local conditions, including also ease of application and safety, packaging and storability and cost when deciding which formulation to use. If tablets are produced locally there may be no choice and cost will generally be the most important factor. Techniques have been developed by the United States Department of Agriculture (USDA) for the fumigation of bulk grain in tankers and bulk dry cargo vessels using phosphine. These techniques are claimed to be both safe and effective under the prevailing circumstances.

The margins for safety and efficacy are very narrow, however, and the techniques should only be used under favourable conditions. Such fumigations may be carried out while the vessel is at sea thus reducing the need to delay shipments. In-transit fumigation of grain has proved particularly valuable to US grain exporters and is now also being carried out on shipments of bulk grain from France, the Federal Republic of Germany and Argentina. The procedure developed by the USDA has been endorsed by the International Maritime Organisation (IMO).

The possibility of some survival of insects during such fumigations was recognised but it has to be accepted that the current inspection and handling procedures often permit no alternative to in-transit ship-board fumigation when cargoes were found to be infested. Ship-board fumigations should be confined to bulk grain in tankers and bulk dry cargo vessels. No techniques have yet been developed for the safe and efficient fumigation of cargo types of ships, break bulk cargoes and tween deckers and fumigation in these vessels should not be attempted. There are three methods by which flat bulks of grain are treated with phosphine

(1) Probing phosphine formulations throughout the bulk with or without covering with gas-proof sheets.

(2) Placing the fumigations on the surface of the grain: fumigation of bulk grain in this way is only satisfactory if carried out at temperatures exceeding 15°C and it is essential that adequate dosages be applied. There is great concern at the low dosages that are now being recomended for use in developing countries. Dosing at levels substantially above the theoretical minimum is needed for full control, especially where there is difficulty in obtaining the high standards of sealing necessary to enable the exposure periods to phosphine required for control.

(3) 'Spot' fumigation, in which only limited portions of the grain bulk are fumigated by probing formulations into areas which are identified as being infested was deplored. Spot fumigation with phosphine has only a 50% chance of success at the first attempt and could lead to a gradual build up of resistance.

Experiments in Togo demonstrated the successful phosphine fumigation of small rural storage structures containing sorghum grain, despite considerable leakage of gas, with application rates of 6 g phosphine/m^3. It was considered that phosphine could be used for the disinfestation of grain developing countries at village or small trader level in appropriate circumstances. The need to ensure safety was emphasised, however, especially as storage installations are often intermingled with living accommodation. This can only be achieved with good sealing which is necessary in any case if efficacy is to be achieved and the danger of build-up of resistance to phosphine is to be avoided. Also there is a need for small packs of phosphine formulations so as to avoid the danger of traders breaking large packs and retailing tablets in unsealed containers.

Sensitive, accurate, reliable, rapid and portable methods of analysis are needed for monitoring concentrations of fumigations in the field. This is necessary both for assurance that adequate concentration – time products are obtained and to avoid exposure of personnel to hazardous concentrations. The most common method used in the field for the rapid determination of phosphine and methyl bromide concentrations is that of detector tubes. These contain cotton wool impregnated with a chemical reagent which changes colour as a gas sample is drawn through the tube.

A measure of the concentration of fumigant is indicated by the length of the decolourised cotton wool. Detector tubes are calibrated directly for this purpose and can be used to give approximate indications of fumigant concentrations within a period of 5 min or so. Detector tubes are not sufficiently accurate for the determination of phosphine concentrations in experimental programmes. Other methods for monitoring fumigant concentrations depend on their physical rather than their chemical properties.

A meter measuring methyl bromide concentrations by thermal conductivity has been in use for 30 years. Portable infra-red gas analysers for the determination of both phosphine and methyl bromide have been used for 10 years or more, while instruments based on sonic analysis or interferometry have also been developed. None of these, however, has found any widespread application except for research purposes. The most accurate method of analysis for both methyl bromide and phosphine remains that of gas liquid chromatography (glc). It is difficult to use this technique in the field, however, except in mobile laboratories. Such facilities are generally only available to sophisticated research organisations.

Two methods for phosphine analysis which have recently been developed were demonstrated. They combine the accuracy of glc with the other requisites of field test equipment. They have significant potential, not only for research purposes but also as a more accurate replacement for detector tubes. The first of these, developed by the Tropical development and Research Institute (TDRI), is based on the reaction of phosphine with mercuric chloride in a modified cell connected to a conductivity meter which provides a direct digital reading of phosphine in parts per million.

The method is simple to operate and has been successfully used in a wide range of field conditions in Africa and Asia. Accuracy, determined by comparison with glc, is about ± 5%, including sampling errors. The equipment may be assembled from standard components, it is cheap and can be carried in a small camera case. The other method developed at Laboratoire Denrées Stockées in France uses the reducing activity of phosphine in an electrochemical cell.

When gas containing phosphine is admitted to the cell an oxydoreduction reaction is induced and produces electric current in proportion to the concentration of phosphine present. This can be converted into a direct reading of phosphine in parts per million on a digital meter. The method has been tested in field trials in France and has provided a reliable method of measuring phosphine concentrations over a range of conditions. It may be possible to adapt it to give a continuous recording of phosphine concentrations.

Phosphine toxicity

Many aspects of phosphine toxicity to insects are still incompletely understood but it is known that oxygen is needed for the gas to be effective and different species, and developmental stages of the same species of insects, vary enormously in their susceptibility. Some stages are so tolerant that they are able to continue normal development in the presence of quite high concentrations of gas. In so doing they pass from the tolerant stage to one of greater susceptibility. If the fumigation area is adequately sealed death of these stages occurs shortly after a susceptible stage is reached.

Obviously, with an actively increasing population comprising adults and immature stages, there is much to be gained if the exposure pelod can be increased to span the tolerant stages in the life cycle. The length of the exposure required is temperature dependent because it is governed by the rate of insect development. Increasing concentration of gas over shorter periods is unlikely to improve control in these circumstances. In tests on adults or diapausing larvae where there is little or no progressive development, the period of exposure to phosphine is still more important concentration.

For larvae in diapause there is much variation in the minimum time of exposure required for death among individuals in tests at high concentrations. Minimum periods to achieve 50 or 90% kills are relatively long compared with other stages. For this reason, efficacy of phosphine, unlike other fumigants is not determined by the product of concentration and period of exposure. Formulae equating concentration of phosphine and exposure periods cannot, therefore, be expressed in terms of the product product producing a constant mortality (as for other fumigants such as methyl bromide). They must also take account of minimum gas concentration and exposure period needed to achieve required levels of mortality.

Superimposed upon the basic toxicity pattern is a narcotic threshold, encountered at higher concentrations; this is a state assumed by insects in which they become dormant and are not killed by phosphine. The concentration range over which this effect is observed may increase at higher temperatures. At concentrations inducing narcosis some insects survive for longer periods than at lower concentrations, but the protection conferred is temporary.

Unexpected survival after relatively long exposures is also encountered when insects experience a gradual rise in gas concentration from zero. This occurs in practice as gas is released from a proprietary formulation. This effect may also be linked with narcosis because, when concentrations rise to 1 or 2 mg/l in about a day, narcotic levels are encountered within a few hours of dosing. At present phosphine toxicity data has largely been obtained using constant concentrations and more work is required on changing levels. The mode action of phosphine is still uncertain, however, some interesting results have been obtained from uptake studies comparing resistant and susceptible strains.

Examples of phosphine resistance so far studied have not been associated with narcosis but physical activity and respiration have appeared unaffected for a substantial initial period of the exposure. Clear evidence has been obtained in Rhyzopertha dominica for a mechanism which actively excludes the gas and which greatly delays the accumulation of poison in the tissues. Four other species have a similar mechanism. Investigations are underway to determine if exposure conditions can be manipulated to overcome this mechanism.

Resistance to phosphine has been detected in insects from a number of developing countries. Tests on immature stages have shown that some resistant strains of several species now require doses as high as the naturally tolerant Sitophilus spp. for control. Resistance has been detected in strains of six species of beetle collected in the UK. It was observed that no discriminating dose test has yet been developed for species of moth infesting stored products.

Work in Australia has shown that resistance can be induced in most strains of insect species by exposure to phosphine in the laboratory. The relative importance of time of exposure and concentration level varied in different strains, although time of exposure was always the more important factor. The minimum level of concentration required for the fumigation to be effective was raised in resistant strains, indicating a greater ability for detoxication. The narcotic threshold was also higher in resistant strains.

No instance has so far been detected in which resistance to phosphine has developed to the extent that infestation cannot be controlled by recommended procedures. However, it is vital that fumigations with phosphine only be carried out where exposure to lethal concentrations can be guaranteed for adequate periods. This implies either good conditions of sealing or the use of novel procedures in which multiple application of fumigant can be carried out. There is also a need to update information on the incidence of resistance in countries where. phosphine is used.

The first positive confirmation of fumigation failure due to phosphine resistance occurred in Bangladesh in 1981. This could be related to poor fumigation practice in which leaky godowns had been fumigated with inadequate doses of phosphine over many years. Substantial levels or resistance were noted in R. dominica, Cryptolestes ferrugineus and Tribolium castanceum. Highly resistant strains of C. ferrugineus have been found on commodities imported into the UK from the Indian subcontinent, suggesting that the resistance noted in Bangladesh was not an isolated occurrence. TDRI is now carrying out a survey of the current geographical incidence of resistance to phosphine.

Resistance has been observed in strains of insects from Pakistan, Nepal, Bhutan, Sri Lanka, Botswana, Mali and Tunisia. Strains from Ethiopia, Liberia, Nigeria,

Zimbabwe, USA, Burma, St. Lucia and Thailand were found to be non-resistant. Evidence was presented to suggest that some of the resistant strains had originated in other countries and had been transferred on imports. This and similar work to update current knowledge of resistance to phosphine is most valuable and should be continued and strongly encouraged. The causes of resistance development have been examined in detail.

The area where resistance appears to be most prominent is in the Indian subcontinent where the fumigation of whole stores without proper sealing is commonly practiced. This is the same cause as the observed resistance in Bangladesh, and fumigation procedures similar to those used in Bangladesh are widely practiced in India, Pakistan and Nepal. Such practices result in repeated underdosing and inadequate exposure periods which cause a gradual increase in the proportion of the total population showing resistance. It was recognised that fumigation of whole stores is likely to continue because the simplicity and cost of the technique made it attractive.

Fumigation of such stores can be improved without any great increase in cost or technical complexity by better sealing, the use of a phosphine formulation which releases gas more slowly and by multiple application of formulation. The need for further research into the development of suitable fumigation techniques, taking more account of meteorological factors such as temperature variation and wind was recognised. The importance of developing such techniques for whole-store fumigation is paramount.

Application of Carbon Dioxide

The use of carbon dioxide as a fumigant for grain stored in large bulks has been demonstrated in the USA,

Australia and other countries. Such treatment is effective if concentrations of about 90% are maintained. For this reason it is likely to be economic only where carbon dioxide is available at very low prices. Interest in this technique is being shown by countries in Southeast Asia where cheap carbon dioxide is available. The use of carbon dioxide mixed with small quantities of phosphine or methyl bromide was described.

Such techniques have been shown to be effective and were likely to be of most use for the fumigation of bulk grain in silos and containers. The use of oxygen-free ('inert') atmospheres to control infestation of bulk grain was described. The atmospheres may be produced by combustion and feeding the exhaust gas into the structure. They have the advantage over carbon dioxide in that they can be readily produced on site and are thus independent of industrial production of gases.

The effective use of inert atmospheres requires a high degree of airtightness as the presence of even very low concentrations of oxygen will allow insects to survive. Such air tightness is most easily obtained in flexible plastic enclosures. Whilst there is potential for such systems there may be difficulties in applying them in most developing countries. Liquid fumigants might be reconsidered to a limited degree as a replacement for phosphine. Potentially they are of particular use for the fumigation of small quantities of bulk grain, spot fumigation in large bulks and the localised fumigation of milling equipment.

Most traditional liquid grain fumigants, e.g. ethylene dibromide or ethylene dichloride administered with carbon tetrachloride, are suspected of possible carcinogenicity and other toxic effects. A new mixture (80% trichloroethane, 20% methyl bromide) was discussed and it was hoped it would meet the new, more stringent, requirements for mammalian toxicity and still be as effective as the former fumigants.

Hydrogen cyanide was at one time extensively used for the fumigation of grain and other commodities but has been almost completely superseded by methyl bromide or phosphine. There might be some scope for the reintroduction of hydrogen cyanide in particular circumstances where phosphine or methyl bromide cannot be used.

—

12

Instructions for All Fumigants

Two Trained Applicators

Label statements that legally require the presence of two trained applicators during hazardous stages of fumigation reinforce long-standing recommendations that fumigators always work in pairs. Most labels require that two applicators work together whenever fumigant application or gas monitoring requires entry into, or work within the confined space where a fumigant is applied.

Aluminum phosphide and methyl bromide labels do allow an applicator to work alone if the fumigant is applied outdoors to a moving grain stream (aluminum phosphide) or in recirculation systems where methyl bromide concentrations do not exceed 5 parts per million in the work area. Even so, the presence of two trained applicators is always a wise investment for safety in the event of accident or emergency.

Gas Monitoring Devices

Revised labels for fumigants require the use of sensitive gas monitoring devices during fumigant application and before warning placards can be removed from fumigated storages. Devices that provide adequate sensitivity include gas detector tubes and matching pumps manufactured by Auer, Draeger, Matheson Kitagawa, MSA, and Sensidyne. Detector tubes are sealed glass tubes filled with a specific reactive solid. Both ends of a tube are broken off just before use, and one end is attached to a calibrated pump.

Available pumps use a bellows, bulb, or piston-type syringe to draw a precise volume of air through the detector tube (Figure 1).

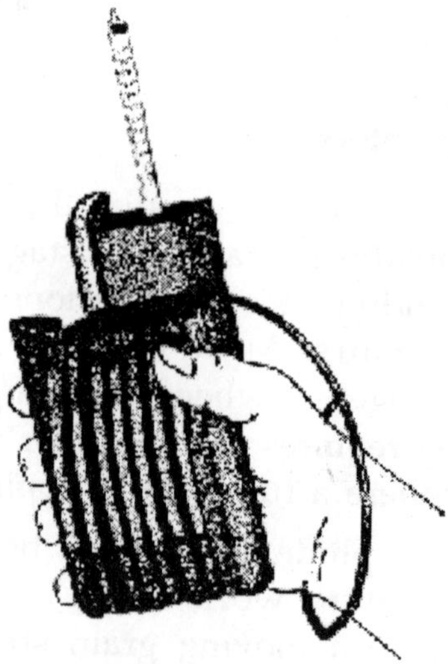

Figure 1. Draeger pump with a gas detector tube attached.

Discoloration of the solid material within the tube indicates fumigant presence, and the gas concentration can be read directly from the scale on the glass tube, as shown in Figure 2. Although tubes and pumps manufactured by different companies may be very similar, accurate readings require matching detector tubes and pumps from the same manufacturer. Do not mix separate brands of equipment.

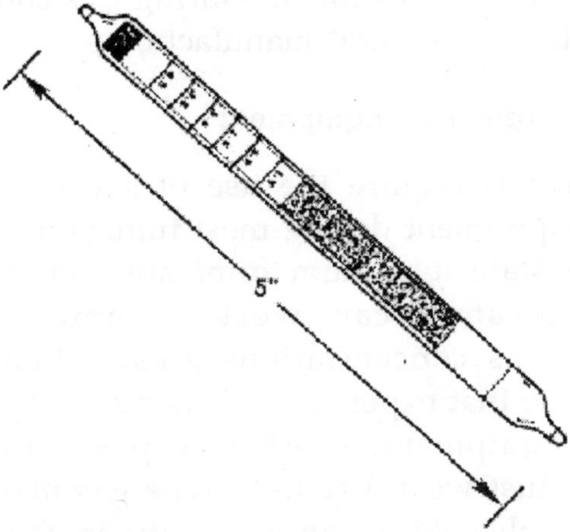

Figure 2. Gas detector tube with a reading of 1.0.

"Low-range" detector tubes that accurately indicate low levels of fumigant concentrations are required for label-specified monitoring practices that provide information for worker safety. "High-range" tubes may be useful for detecting fumigant leaks. These tubes are scaled for measuring much higher concentrations of fumigants, and they are especially useful for monitoring gas concentrations within storages during fumigation to determine whether necessary fumigant levels have been reached.

Other gas monitoring devices such as halide leak detectors and thermal conductivity meters may be used to detect leaks or determine internal concentrations of gas during fumigation, but these devices do not provide label-required levels of sensitivity necessary for determining safety needs such as respiratory protection. Detector tubes are specific for a single fumigant. Auer, Draeger, Matheson-Kitagawa, MSA, and Sensidyne manufacture detectors that offer adequate sensitivity for label-required monitoring of hydrogen phosphide and methyl bromide. Tubes for measuring CO_2 concentrations are available from several manufacturers.

Respiratory Protection Equipment

Fumigant labels require the use of specific respiratory protection equipment during most fumigant applications. Labels also state maximum fumigant concentrations in which applicators can work without respiratory protection. Gas concentrations greater than specified levels indicate that exposed workers must use respiratory protection equipment. Workers exposed to hydrogen phosphide must wear a canister-type gas mask or SCBA if hydrogen phosphide concentrations in the work area exceed 0.3 parts per million.

At concentrations greater than 15 parts per million or when hydrogen phosphide gas concentrations are not measured, workers must wear a SCBA. Workers exposed to chloropicrin concentrations greater than 0.1 part per million must wear a canister-type respirator, a SCBA, or a combination air-supplied-SCBA respirator. Workers exposed to methyl bromide levels greater than 5 parts per million must wear a SCBA or a combination air-supplied-SCBA respirator; and a SCBA or combination air-supplied-SCBA respirator is required when CO_2 concentrations in work areas exceed 1.0 percent.

Canister and cartridge-type gas masks or respirators employ a replaceable canister or cartridge containing chemical components that absorb specific gases. Full-face, canister respirators (not half-face cartridge respirators) should be used when this type of respiratory protection meets label specifications. The effective life of an individual canister varies according to fumigant concentration and the respiratory rate of the applicator.

Maximum limits are stated on each canister. Under NIOSH/MSHA regulations, canisters are colour coded according to fumigant absorbency. Canisters approved to provide protection from hydrogen phosphide are coded yellow with an orange stripe; canisters effective for protection against chloropicrin are colour coded black. No canisters are currently approved for protection against methyl bromide. In addition to checking colour coding, applicators must be certain that written specifications indicate that a canister is effective against the fumigant to be used. When a canister is no longer effective, it becomes hot.

Some canisters have a colour gauge or "window" that shows when they are about to become ineffective. Breathing hot air or encountering high resistance to breathing are also warnings that the canister is about to become ineffective. When this occurs, or when an applicator smells or tastes the fumigant or becomes nauseous or dizzy, he or she should exit the fumigated area immediately. An expired canister should be crushed before it is discarded so that no one will mistakenly use it in the future.

Canister respirators are inadequate for use in oxygen-deficient environments such as those produced during CO_2 fumigations. Although the canister may absorb the toxic fumigant in this environment, it cannot supply necessary oxygen. In these situations, a SCBA or a combination air-supplied-SCBA respirator is needed. The SCBA commonly used for fumigation is the air pack (Figure 3). Air packs consist of a full-face mask attached to a tank of air carried on the applicator's back. An air pack supplies up to 25 to 30 minutes of air and allows work in an oxygen-deficient environment. This time period may be considerably shorter if overexertion increases the rate of respiration.

A warning bell can be set to signal the depletion of the air supply. In a fumigant-laden storage, safe exit may require uninterrupted respiratory protection. For this reason, carrying an approved canister respirator when using an air pack is recommended where oxygen concentrations remain adequate. The canister respirator will allow emergency escape if the SCBA expires or malfunctions.

Figure 3. Applicator wearing an air pack (SCBA) and using a piston-type gas detector pump.

Methyl bromide label directions concerning respiratory protection include reference to combination air-supplied-SCBA respirators. Air-supplied respirators employ an outside air source pumped to the applicator through an air line. An air line system has the major advantage that the air supply does not expire in a short time. Disadvantages, however, include: the airline must be towed throughout the storage; air pump failure or a constriction of the air line can shut off the air supply; and the air pump must be located in a fumigant-free area.

In combination with a SCBA, an air-supplied respirator does offer an unlimited work period with backup respiratory protection provided by the SCBA if for any reason the outside air supply is cut off. One final respiratory protection topic concerns the fit of face masks. A face mask that does not seal tightly against the face cannot provide protection from a fumigant gas. An applicator must select a mask that fits his or her face; facial hair must be cleanly shaved to allow a tight fit. Fit can be tested by closing off the breathing tube and attempting to breath in or blow out air.

Passage of air between the mask and the face indicates an unsatisfactory fit. Another test involves the release of an irritant gas in front of the mask. If the gas leaks into the mask, the wearer will experience eye or respiratory irritation or nausea-indicating a poor seal.

Respiratory protective equipment is of two basic types:

1. Supplied-air respirators may supply uncontaminated air from outside to the mask or carry the air with them (self contained bang apparatus).
2. Air-purifying respirators (gas masks and edge respirators) purify the air by removing gasses and particles from the air.

Individuals who have difficulty breathing with air purifying respirators should use a respirator which supplies air itself or from the outside. You should use an air-supply mask if you suspect exposure concentrations exceeding the concentration or time limitations of the gas mask's canister. Individuals who have punctured eardrums may draw in fumigant vapours through the ear as a result of creating a slight negative pressure during inhalation. Any fumigant drawn in this way will be exhaled into the respirator face piece and a poisonous concentration may build up inside the respirator.

Persons with punctured ear drums can usually obtain complete protection by using cotton earplugs smeared with oil. Individuals wearing respirators as part of their work should have physical examinations to make certain they are physically fit to wear respirators safely.

Supplied-Air Respirators

These respirators supply fresh clean air from an outside source or cylinder rather than purifying air at the location of use. They should have a blower or positive pressure to ensure an air supply. Supplied-air respirators are required when oxygen has been depleted or the concentration of fumigant is unknown These respirators are more expensive than canister respirators, but the latter have several disadvantages:

- Canisters have limited life and must be replaced.
- Special canisters must be available for each fumigant.
- The canister provides no protection at abnormally high fumigant concentrations or in spaces where oxygen is deficient.
- Air-supply respirators, like gas masks, should be properly fitted and used according to instructions.

Air-Line Masks

These devices have face pieces similar to the canister gas masks except that the masks are attached to hoses which lead to an air pump. The air pump is usually powered by a gasoline engine which forces air through a filter. One pump can supply air for several lines. This respirator allows you to operate for an indefinite time within a space being fumigated. (Note: These respirators provide no protection against skin absorption of the fumigant). Disadvantages include:

- The fumigator's movement is somewhat restricted as he must tow the air line;
- The air line can become caught or kinked, shutting off the air supply;
- The engine may fail;
- The-air pump must be located so that contaminated air is not pumped to the fumigator;
- There is a 250 ft limit on hose length;
- Care should be taken to avoid cutting or damaging the hose.

Self-contained Breathing Apparatus

A self-contained breathing apparatus (SCBA or air pack) consists of a full-face mask attached to a tank of air carried on the back of the fumigator. It gives the fumigator the mobility of the canister mask, and yet does not tie him to an air pump. With this device, the fumigator can work in any gas (provided additional protective equipment is worn to prevent dermal exposure). With the popularity of diving, it is not difficult to get the air tanks refilled. Depending upon the size of the tank, the air supply will last up to 30 minutes. There is usually a warning bell to alert the fumigator when his air supply is running low. The main disadvantage of the air pack is that the fumigator has to carry a heavy tank while he complete his work.

Air-Purifying Respirators

Air-purifying respirators are designed to remove particulate matter and/or toxic vapours from the air. Their capacity to purify air is limited by the size of the cartridge or canister, the concentration of toxic vapours, and the ability of the contents of the cartridge or canister to remove the toxic substances.

Chemical Cartridge Respirator

These respirators, in addition to being equipped with filters to remove fine particles, also have one or two chemical cartridges which remove toxic gasses and vapours, either chemically or physically by absorption. The sorbent volume is small and the cartridge's lifetime is generally short. Thus, these respirators are used for low concentrations and short exposure periods. Cartridge respirators are not suitable for use with fumigants. The cartridges are too small and they tend to leak around the face piece.

Gas Mask

These respirators are equipped with canisters containing sorbent for removing toxic gasses and a filter for removing particles. They contain more sorbent than cartridge respirators, and may be used for higher toxicant concentrations and/or longer exposure periods. These respirators are suitable for use with fumigants. However, the canisters must be suitable for protection against the fumigant being used. Because gas masks are air-purifying devices, designed solely to remove specific contaminants from the air, their use must be restricted to situations in which the air contains at least 19.5% oxygen by volume and, generally, no more than 2% toxic gasses and vapours by volume.

If the oxygen content or concentration of fumigant is not known, measure them with instruments (such as gas detector tubes) to make certain the correct canister and respirator is used. The life of the canister is limited and varies with the fumigant used and the fumigant concentration The maximum permissible limits are usually stated on each canister. Do not exceed these maximum limits.

Canisters

For most gas masks, canisters are available in four sizes. The chin-size is the smallest canister and should not be used in fumigation work because of its low capacity and concentration limitations. The industrial-size canister is the most commonly used canister in fumigation, but many people recommend using the super-size canister because of its greater capacity. The largest canister, the N size, has about twice the capacity of the industrial-size. You should select a canister that is produced by the same manufacturer as is the mask that you plan to use.

The period of safe use of the canister depends on a number of variables, including:

- The canister type,
- The canister size,
- The kind and concentration of fumigant(s),
- The length of exposure period,
- The rate of breathing,
- Whether or not there is more than one gas present, and
- The temperature and humidity at the time of use.

Never use a canister after the expiration date listed on it. The gas mask, when properly assembled and fitted, is a compact air-purifying unit which finishes protection against those harmful vapours listed on the canister label. During inhalation, the air enters the canister through the bottom and passes into the interior. Here the air is chemically purified or the harmful vapours neutralised. The purified air then passes through the corrugated rubber tubing into the molded channel of the face piece. Some of these channels direct the purified air to the lenses to reduce fogging.

During exhalation, the air is expelled from the face piece through the exhalation valve, which is designed to permit near-normal conversation. This valve outlet also serves as a drain for moisture which might otherwise condense within the face piece as the operator breathes. An inhalation valve at the bottom of the canister prevents the exhaled air from being expelled through the canister.

Each person should have his or her own respirator and canister; canisters should not be shared with others. Canisters should not be reused; when a canister is no longer useful or is being discarded you should destroy or mutilate the top so that no one else can use it. The respiratory mask should be cleaned and disinfected after each day's use and at least once a month. If you do not sanitise the mask immediately, you should at least wipe out the interior with a clean cloth, preferably one saturated with isopropyl alcohol.

To sanitise masks, prepare a solution of cleaner-sanitiser and warm water, and immerse the mask in this solution. The interior and exterior of the mask should be scrubbed with the cleaning solution and a sponge. The mask should be thoroughly rinsed with warm water and be allowed to dry. During cleaning, inspect the mask for rubber deterioration and tightness of connections, and the face piece for a tight seal. Keep a record of these cleanings and inspections.

After cleaning and inspecting the mask, place it in its carrying case. When not in use, the mask should always be kept in the case to protect it against dust, sunlight, heat, extreme cold, moisture, or damaging chemicals. Experienced persons should make replacements or repairs only with parts designed for that particular respirator.

The respirator must fit well to effectively protect the wearer. A cleanly shaven face and short sideburns make a good fit easier. Insert your chin into the lower part of the face piece and pull the headbands back over your head. To obtain a firm and comfortable fit against the face at all points, adjust the headbands as follows:

1. Straps should lie flat against your head.
2. Tighten the lower, or neck, straps.
3. Tighten the side straps.
4. Place both hands on the headband pad and push it toward your neck
5. Repeat steps 1 and 2.
6. Tighten front, or forehead, strap a few notches.

Before using the mask, test that it is tight by squeezing or pinching off the corrugated breathing tube, or by holding your hand tightly over the inlet end of the canister. Inhale gently so that the face piece collapses slightly and hold your breath for 10 seconds. If the mask is tight, it will remain collapsed while you hold your breath. If any leakage is detected around the face seal, readjust the head harness straps and repeat the test until there is no leakage. The face piece must pass the tightness or "good fit" test before you enter a toxic atmosphere.

You will not be protected unless all of the air is drawn through a suitable canister or from a source of clean air. If you cannot find a leak in the face piece, check the hoses and connections to make sure they are tight and in good condition. A new rubber washer for the mask hose is supplied with each new canister. This washer must be in place when attaching the hose to the canister, or vapours could enter at this point; be sure and check for this washer. If a leak still exists, try a new corrugated breathing tube. If this takes care of the leak, destroy the defective breathing tube.

If, after removal of the seal or your hand from the canister inlet, you find you cannot breathe, the canister must have a blockage. Destroy and replace it. If the respirator is a supplied-air type, check the face piece and breathing tube as described for the gas mask. If you are using a self-contained breathing apparatus, check the air tank for leakages, amount of air, and valve efficiency. Test the valves, connections, and hose on supplied-air respirators.

The respirator should be ready for use at all times and, when not being worn on a job, should be on hand for emergency use. The following is a checklist for proper use of your respirator:

1. Determine the type of hazard. If the air is deficient in oxygen (less than 19-5% by volume), do not use a gas mask and canister: use a supplied-air respirator. If in doubt, always use a supplied-air respirator.
2. Remove the respirator from its case and check all rubber parts and fittings for deterioration and a tight fit.
3. If you use a gas mask, check the canister for its expiration date. Always use a new canister.
4. Select the proper canister for the gas that is to be used. Make sure that it is the proper canister by reading the label carefully, and be sure it is the correct canister for the respirator you are using.

5. Put in the new washer that comes with each canister.
6. Remove the tape that covers the intake port at the bottom of the canister.
7. Connect the mask and canister.
8. Put the mask on in fresh air.
9. Check for proper fit and leaks.
10. Note the current time and the time when you should be out of the fumigated area.
11. Enter the contaminated area cautiously. Return to fresh air immediately if irritating gasses, odors, or symptoms of distress are noticed.
12. After completing the job:
    - Clean and inspect the respirator.
    - Record the date of cleaning.
    - Mutilate top of the canister so that it cannot be used and discard it.
    - Return the respirator to its carrying case and place it in a storage area.

Importance of Warning Placards

Labels specify the wording and content that must appear on warning placards (including the presentation of some information in both English and Spanish). These requirements represent no major changes in a fumigator's procedures. Regulations do specify an important step concerning the removal of warning placards. Following aeration, warning placards may not be removed and the commodity may not be processed or fed until a certified applicator uses an appropriate gas detection device to determine that gas concentrations have dropped below levels specified for each fumigant. Such follow-up monitoring was not previously conducted by many applicators.

Special Caring for Fumigant

Fumigants are hazardous materials; empty containers and any excess or unused fumigant must be handled appropriately. The correct method of handling containers differs among fumigant products. Empty canisters or tanks that were used to hold methyl bromide or CO2 under pressure should be returned through the original shipper to the manufacturer for refilling. Always consult specific container labels for instructions.

Unused aluminum phosphide pellets or tablets in opened flasks should not be disposed of. If flasks are resealed, these tablets or pellets can be stored safely for future use as long as the label remains intact. Do not store flasks at sub-zero temperatures, because doing so will increase the likelihood of ignition when they are opened. If aluminum phosphide tablets or pellets are spilled or flasks are punctured, hydrogen phosphide gas is released.

Persons cleaning up the spill or working in the contaminated area must wear a SCBA unless gas detection equipment is used. If gas concentrations are measured, a canister respirator should be used if hydrogen phosphide concentrations range between 0.3 and 15 parts per million. At higher concentrations a SCBA is required. Technically, the full-face canister respirator can be used in concentrations up to 1,500 parts per million for emergency escape purposes, but not to re-enter an area where concentrations exceed 15 parts per million.

Cleanup personnel should wear cotton or neoprene gloves while handling spilled material. If a spill can be cleaned up immediately, spilled pellets or tablets should be used immediately or transferred to an empty flask with an intact label. If such a container is not available, tablets or pellets can be placed in a sound, dry metal container, which should be sealed and labelled as aluminum phosphide. Keep the original product label with the substitute container.

If spilled material has begun to react or if it is contaminated by other substances so that it cannot be safely stored, it should be placed into open-top, perforated gallon cans and processed immediately. Do not use water to clean up an aluminum phosphide spill. Water will react with tablets or pellets to release hydrogen phosphide gas, the rapid production of gas can cause spontaneous ignition and explosion. To deactivate unreacted or partially reacted pellets, they should be transported by hand or in an open vehicle to open air away from occupied structures.

Fill a drum two-thirds with water and add one-quarter cup of low-sudsing liquid detergent or surfactant for each gallon of water. Each flask of tablets or pellets should be mixed with no less than 1-gallon of the water-detergent mixture. Wear respiratory protection equipment and slowly add the aluminum phosphide product to the drum while stirring. Stir occasionally thereafter for at least 36 hours.

Do not cover the container! Covering the container will confine the hydrogen phosphide gas that is generated, and the resulting high concentrations may explode. This wet method of deactivation is preferred when five or more flasks of material must be deactivated. The resulting slurry may be disposed of at an approved landfill or burial site. An alternative to slurry deactivation is dry deactivation (for quantities not exceeding five flasks). Pellets or tablets can be spread out in an open, secure area away from occupied buildings and deactivated by atmospheric moisture.

Disposal of residual dust from reacted pellets or tablets is necessary following a space fumigation. Residual dust is grayish white, and it contains a small amount of unreacted aluminum phosphide. (Tablets or pellets that are only partially reacted remain slightly greenish in colour; they should be disposed of in the manners described above for spills.) Residual dust from up to five flasks can be disposed of by on-site burial or by spreading over the land surface in a secure area away from inhabited buildings. This amount of dust may also be disposed of at a sanitary landfill or approved pesticide incinerator.

For larger amounts of residual dust, a detergent slurry disposal method similar to that described above is recommended. Residual dust from up to three flasks can be held in an open one-gallon bucket pending disposal. Larger amounts of dust should be held in a porous cloth bag during any storage or transport before disposal. Do not put the residual dust from more than eight flasks of tablets or ten flasks of pellets in any one bag before disposal. Greater amounts may generate enough gas that a risk of explosion may exist.

Do not pile these bags or leave them exposed to rain. Do not confine, dispose of, or store residual dust in closed containers such as dumpsters, drums, or plastic bags. Do not dispose of dust in toilets. Empty flasks that contained aluminum phosphide may be recycled or disposed of in sanitary landfills after they have been properly processed. To clean flasks adequately for to disposal, flasks and stoppers may be triple-rinsed and then punctured, or a small number of empty flasks may be punctured and held outdoors in an open and secure area away from occupied buildings to allow complete reaction of aluminum phosphide. Where triple-rinsing is used, rinsate may be disposed of in a sanitary landfill.

Common Sense

It is essential that fumigators understand and follow the technical instructions that promote safe and effective fumigation of stored grain. It is just as important that fumigators remember to use good common sense when planning and carrying out a fumigation. Although it may be impossible to teach good common sense by writing instructions in study materials, the following comments are offered as reminders to exercise good judgment and think ahead.

- Read and understand label directions. Demand information from the manufacturer, distributor, and Extension personnel. Do not use a fumigant without adequate training and confidence in your ability to do the job properly.
- Supply local medical personnel with fumigant and poison treatment information before using the fumigant.
- Preplan the entire job. Think through every step, and plan your reactions to possible problems and emergencies.
- Always work in pairs.
- Use or have available proper safety equipment. Make sure all equipment fits well and that all applicators are trained in and familiar with the use of necessary safety equipment.
- Do not take short cuts; follow through with well-planned and thorough application practices.

- Do not become complacent. Each job is a new challenge and a new situation in which an emergency may require rapid and proper reaction.
- One phrase may be appropriate: "Do not be the fastest fumigator; be the oldest."

Bibliography

Adomako, D., "Radiotracer study of methyl bromide residues in fumigated cocoa beans", *Proc. 4th Conference of West African Cocoa Entomologists*, 1974.

Agarwala, S.B.D., "Melanaspis glomerata (Greur) - a new Coccid pest of sugarcane", *Proc. Biher Acad. Aqric. Sci.*, 5, 1956.

Aharoni, Y., Stewart, J.K., Hartsell, P.L. & Young, D.K., "Acetaldehyde a potential fumigant for control of the Green Peach Aphid on harvested head lettuce", J. Econ. Entomol., 1979.

Beran, F., "The effect of various fumigants on San Jose scale", *Gen. Rep.* 1st int. Congr. Plant Prot., 1946.

Berck, B., "Distribution and persistence of fumigant mixtures applied to grain", *Proc. 10th int. Congr. Ent.*,1958.

Caylor, J.F. & Laurent, C.K., "The effect of a grain fumigant on the egg size of hens", *Poult. Sci.*, 1960.

Childs, D.P. & Overby, J.E., "Atmospheric chamber fumigation of cigar tobacco", *T . Sci.*, 1967.

Cogburn, R.T. & Tilton, E.W., "Studies of phosphine as a fumigant for sacked rice under gas-tight tarpaulins", *J. Econ. Entomol.*, 1963.

Gammon, E.T., "Methyl bromide fumigation of camelia cuttings", *Bull.* Calif. Dent Aqric., 1950.

Greve, P.A. & Hodgendoorn, E.A., "Determination of fumigant residues in grain", *Meded. Fac.*, Landbouwwet. Rijksuniv. Gent., 1979.

Halliday, D. & Prevett, P.F., "Fumigation of pyramidal stacks of bagged decorticated groundouts with methyl bromide", *J. Sci. Food Aqric.*, 1963.

Heseltine, H.K., "The use of thermal conductivity meters in fumigation research and control", London, Agricultural Research Council. *Pest Infestation Research Bulletin,* 1961

Joffe, A. & Nolte, M.C.A., "Methyl bromide fumigation of horizontally stored bulk maize", *J. Entomol. Soc. S.* Afr., 1957

Kenaga, E.E., "Evaluation of grain fumigants", *Down to Earth,* 1957.

Leesch, J.G. & Highland, H.A., "Fumigation of shrink-wrapped pallets", *J. Georgia Entomol. Soc.,* 1978.

Lindgren, D.L. & Vincent, L.E., "Fumigation of food commodities for insect control", *Adv. Pest Control Res.,* 1962.

Rai. L., Sarid, J.N. & Pingale, S.V., "Fumigation of food grains in India with hydrogen phosphide", *I. Tests in concrete bins. Bull. Grain Technol.,* Hapur, U.P., 1963.

Index